Soil conservatic
for small farmer
in the humid tropics

by
T.C. Sheng
Professor
Colorado State University
USA

Soil Resources, Management
and Conservation Service
FAO Land and Water Development Division

FAO
SOILS
LETIN
60

FOOD
AND
AGRICULTURE
ORGANIZATION
OF THE
UNITED NATIONS
Rome, 1989

M-57
ISBN 92-5-102869-9

FOREWORD

Our existence depends laregly upon the thin layer of soil from which we produce most of our food, fibres and timber. If used for the right purpose, and managed well, soils can continue to produce and to provide for our needs indefinitely. But if put to the wrong use, and badly managed, soils can quickly degrade, decline in fertility and lose their potential to provide us with the things we need.

Nowhere is this more so than in the humid tropics. In this region soils tend to contain most of their fertility in the top few centimetres of their profile. If these soils are exposed, and left unprotected from the frequent, high-intensity rainstorms which are common in the humid tropics, they can quickly erode, losing their fertile topsoil and leaving behind poor, infertile land.

The humid tropics are important particularly as they are the home for many millions of small farmers. Most of these farmers are poor and many of them are forced to farm small plots of land on steep hillsides where the risk of erosion is at its greatest. For these people, the control of soil erosion is of the utmost importance - if their land is allowed to erode its fertility quickly falls, crop yields decline and the farmer may well be faced with the prospect of starvation or of migration to the slums of a city in search of work.

In spite of the need, surprisingly few textbooks or manuals have been written specifically for the humid tropics, to help those people involved in practical problems of arresting soil erosion in this part of the world. FAO has therefore produced this Soils Bulletin as a reference for the planners and technicians working with small farmers in the humid tropics and who are searching for ideas and guidance in their efforts to overcome soil erosion and to introduce sustainable systems of productive agriculture.

CONTENTS

LIST OF FIGURES

LIST OF PLATES

LIST OF TABLES

1. INTRODUCTION

1.1 OBJECTIVE

This bulletin describes problems, approaches and techniques of soil conservation in the humid tropics and in particular is intended to assist farmers on small holdings in these regions ('small farmers') to overcome their soil erosion problems. The terms which frame this objective are defined briefly as follows:

- Soil Conservation: The scientific use and protection of land; including wise choice of land use and the pursuit of necessary measures of soil management and of erosion control (especially against erosion by water).

- Small Farmer: Any farmer who farms less than 5 hectares. An average would be around 2 hectares.

- Humid Tropics: Land between the Tropics of Cancer and of Capricorn having a tropical climate and annual rainfall of at least 1 000 mm.

1.2 SOIL CONSERVATION NEEDS

1.2.1 Soil Vulnerability

Soils are the very basis of our existence. Through the past, in the present, and through the foreseeable future, they remain the foundation of our food supply chain - a vital recurrent and capital resource of each nation. People should be keenly aware that the soil mantle which supports human life is very thin and that soil formation is a slow process. Once the thin top layer is eroded away it is difficult to restore. Damage invisible to the naked eye may seriously affect productivity. Soils are much more vulnerable than is generally thought. Only under proper management can they be regarded as renewable resources.

In the humid tropics, where many of the developing countries are situated and where individual holdings are usually small, the risk of soil erosion is likely to be high due to frequent and intense rains. When exposed by improper farming and cultivation the soils in these areas can be badly eroded in a short time. The need for careful soil conservation in these areas is apparent.

1.2.2 Historical Lessons

History shows us that neglect and abuse of soil resources has led, in many instances, to human suffering and even to the downfall of countries and civilisations. Human tragedy repeated periodically on the banks of the Yellow River - 'China's sorrow' - provides a well-known example of suffering caused by mis-use of a watershed. In many countries, from Asia to the Near East and North Africa, once green and productive lands have become barren deserts through the abuse of soil resources. Lands described in the Bible as 'flowing with milk and honey' three thousand years ago are now badly eroded, leaving bare hills and bedrock. Countries which were the granaries of the Roman Empire present the same picture. Dr. Walter Lowdermilk, one of the pioneers of soil conservation in the USA, has described these changes in detail in his book 'Conquest of the Land through 7,000 Years'. He suggested a need for an Eleventh Commandment to safeguard the land 'that thy descendents may have abundance forever'.

1.2.3 Pressing World Problems

i. Land degradation

A study by FAO and UNEP concludes that between 5 and 7 million hectares of land worldwide are lost to production annually through soil degradation (FAO/UNEP

Plate 1 Shifting cultivation of yams on sloping land causing severe erosion (see Plate 2)

Plate 2 After bench terracing, the same land can be used for settled and continuous farming

1983). If current rates of land degradation continue, close to one third of the world's arable land will be destroyed by the year 2000 (UN 1978). Of all the forms of land degradation, soil erosion is the major problem. FAO further pointed out that between 1980 and 2000 the area of arable land available **per capita** in developing countries is set to fall from 0.37 to 0.25 ha with a 19 percent loss of overall land productivity (FAO 1981, 1984a).

ii. Population increase

Regardless of major efforts to encourage birth control, world population is predicted to increase from about 5 billion at present to 6 billion by the year 2000. Most of the increase will be in developing countries and even more pressure is likely to be placed on land which is already degrading.

iii. Food supplies

FAO has estimated that the number of severely undernourished people in the developing world increased from about 360 million in 1969-71 to about 435 million in 1974-76, an increase of 20 percent in only 5 years. The 1985 famine in Ethiopia and other African countries has vividly illustrated the seriousness of the problem. Food and other agricultural production levels need to be doubled in developing countries within the next 15 years merely to keep pace with population growth (FAO 1981). Some developed countries have capacity for increased production but in supplying the needs of developing countries there are attendant problems of transportation and distribution. Accordingly, 'self-sufficiency' is probably a wise agricultural policy for many developing countries to pursue but, as more land is brought under food production, greater efforts will be needed to preserve the soil.

1.3 BENEFITS OF SOIL CONSERVATION

A better appreciation of the value of soil conservation is obtained by recognizing two categories of benefit; on-site benefits and off-site benefits. Many people equate soil conservation with erosion control. Certainly, measures that minimize erosion and reduce sedimentation and flood damage are very important but, in developing countries especially, soil conservation programmes can generate other major benefits which are often overlooked. These other benefits are outlined in the following paragraphs.

i. Inducing permanent farming

Given proper soil conservation and management, many areas could be farmed permanently and much more intensively without risking undue erosion. For instance, the construction of bench terraces would permit settled, continuous farming (Plate 2) in many areas where shifting cultivation (Plate 1) is presently being practised. Shifting cultivation currently occupies 300 million ha of land, mostly in developing countries (FAO 1978).

ii. Increasing the population supporting capacity of the land

Well-planned soil conservation introduces a better choice of land use, improved farming practices, conservation of soil moisture and other measures designed to increase agricultural production and thus raise the population supporting capacity of the land. In some developing countries, cultivated land can support only 2.5 to 5 people per hectare (1 to 2 people per acre). In contrast, some Asian countries support up to 15 people on each hectare of their terraced paddy lands. Increase of population supporting capacity to this level will rarely be feasible but any increase can be important in densely populated countries with limited land resources.

iii. Developing new land safely

FAO has estimated that to feed the world in the year 2000 an additional 150 to

200 million ha of new land should be brought into production (FAO 1981). This expansion will inevitably embrace land inherently less favourable for farming, such as more steeply sloping land in the humid tropics. Sound soil conservation measures will be essential to the safe development of these kinds of land on a sustained basis.

iv. Modernizing upland farming

More intensive use of hilly uplands in many countries seems inevitable. It may also be desirable. In the tropics where the lowlands are hot, the uplands and hill regions often offer climatically ideal places to produce many high value crops for domestic use and export. Upland farming in many countries remains primitive because the hilly terrain is rough and remote. Properly designed conservation structures in these areas will not only protect the hillslopes but will also provide better access (e.g. using terraces for road alignment), better drainage (e.g. waterways) and increased potential for mechanization and irrigation.

v. Providing employment opportunities

Soil conservation and land use intensification activities could create much-needed employment opportunities in the rural areas of many developing countries. Most traditional soil conservation practices are highly labour-intensive. Four to five hundred person-days of work may be employed in constructing one hectare of bench terraces on a moderate slope. Even simple soil conservation structures may require 60 to 80 person-days per hectare under similar conditions. These estimates exclude work on waterways, gully control, etc. Thus a substantial national soil conservation programme with government financial support could effectively alleviate rural unemployment problems prevalent in many developing countries.

2. PROBLEMS

2.1 PROBLEMS RELATED TO NATURAL CONDITIONS

2.1.1 Rainfall and Climate

Rates of soil erosion are closely related to the amount and intensity of local rainfall. Large amounts of rain are likely to saturate the soil and then produce high rates of runoff, whilst high intensity rain exerts a great disruptive and detaching force on the soil particles. Both the intensity and amount of rain can be very high in the humid tropics. Records in Ibadan, Nigeria, for instance, showed that no fewer than 47 storms in 1972 had an intensity higher than 75 mm/h (Lal 1974). In northwest Jamaica between 1973 and 1976 there were 7 events in which 30 minute rainfall intensities exceeded 75 mm/h. In 3 of these the intensity exceeded 100 mm/h. Table 1 compares 'rainfall erosivity' (a calculated value reflecting intensity and frequency of rains) in the humid tropics with that of some temperate countries.

Table 1 RAINFALL EROSIVITY IN SELECTED COUNTRIES (from Posner 1981 and Huang 1976)

Location	Erosivity*	Source
Humid Tropics		
Jamaica - Smithfield	2 603	Hutchingson & Forsythe (1975)
Peru - Iquitos	2 600	Paulet (1978)
Côte d'Ivoire - Abidjan	2 192	Roose (1977)
Taiwan - Southwest	1 534	Huang (1976)
Dominican Republic - Quemados	1 428	Paulet (1978)
Puerto Rico - Mayaquez	1 353	Barnett (1971)
Temperate Zone		
USA - Illinois	350	Wischmeier & Smith (1978)
USA - Gulf Coast	950	Wischmeier & Smith (1978)
France - Central region	100-600	Roose (1977)

* R factor used in the Universal Soil Loss Equation (USLE)

In Southeast Asia heavy falls of rain are concentrated in the 4 to 5 months of summer and autumn when typhoons are frequent. During Typhoon Gloria (10-11 September 1963), for example, the rainfall recorded in 24 hours at one of the hill stations in northern Taiwan was 1 248 mm - more than the annual rainfall in many parts of the world.

Ironically, in many parts of the humid tropics the months which receive characteristically large amounts of high intensity rainfall alternate with several consecutive months of intense dry season. The wet months call for conservation measures focussed primarily on safe runoff disposal. In the remainder of the year soil moisture conservation, water harvesting, or even small-scale irrigation may be the primary need because in the dry months high temperatures and high evaporation severely reduce soil moisture content and limit possibilities of vegetative growth. Together, these contrasting conditions create conservation problems quite different from those of other zones.

2.1.2 Topography

In the humid tropics, large numbers of small farmers cultivate foothills, uplands and mountain sides. Typically, the more fertile plains and alluvial valleys are occupied by the big properties and plantations leaving the surrounding, more hilly lands to the small subsistance farmers. In most places the traditional farming system for these small farmers has been swidden or shifting cultivation. Under this system, a farmer would cut and burn a plot of forest and then cultivate the plot for perhaps two or three years. After this time, the soil fertility would decline, weeds would become a problem and yields would drop. The farmer would then abandon this plot and clear a new one, the old plot being left for perhaps ten to twenty years under a 'bush fallow' before being cleared and cultivated again. While population densities were low, and there was plenty of land, this system worked well; the fertilty of the land was replenished under the fallow period and little lasting damage was done. But in most parts of the humid tropics, population numbers have increased rapidly in recent years and there is too much demand on the land for it to be left for long periods under fallow. Bush fallow periods are therefore either being greatly reduced in length or the land is being placed under permanent cultivation. Under these conditions, the soil does not have a chance to regain its fertility, the soil structure breaks down and soil erosion rapidly carries away the more fertile top soil.

As previously stated, shifting cultivation is being carried out by small farmers on 300 million hectares of the humid tropics. These lands are agriculturally marginal and, for the most part, sloping but they produce basic food crops for more than 250 million people (Bishop 1982). These figures are estimates but could be expected to double by the year 2000 (Bene **et al.** 1977).

A detailed study financed by the Rockefeller Foundation (Posner and McPherson 1981) has illustrated the importance of steeply sloping lands to the national economy of many tropical American countries. Data from this study, given in Table 2, show that such lands occupy between 45 and 80 percent of the countries concerned, and support between 20 and 65 percent of the agricultural populations.

Table 2 PROPORTIONS OF NATIONAL AREA, ARABLE LAND AND POPULATION ON STEEP SLOPES OF TROPICAL AMERICA (ESTIMATED PERCENTAGES) (from Posner 1981)

Country	National Area	Arable Land*	National Population	Agricultural Population
Colombia	40	25	15	50
Costa Rica	70	25	20	30
Dominican Republic	80	15	15	30
Ecuador	65	25	25	40
El Salvador	75	40	30	50
Guatemala	75	30	40	65
Haiti	80	70	50	65
Honduras	80	15	15	20
Jamaica	75	50	15	30
Mexico	45	20	15	45
Panama	80	10	15	30
Peru	50	25	25	50

* Arable land includes only the land used for annual crops. It refers to cropped and/or fallow land which is part of the normal rotation. Thus, arable land includes all the land in sugarcane, cotton and other annual crops, but excludes perennial crops such as coffee and bananas and permanent pasture lands.

Despite their importance to national economies, slopelands and uplands in the humid tropics are often improperly used. In terms of development and protection they receive far less than their share of national and international attention. The entire problem is often left to the small farmer. In many countries, there is a pressing need to pay more attention to the development and protection of uplands.

2.1.3 Soils, Erosion and Runoff

i. Soils

Some soil scientists apparently believe that the kinds of soil most commonly found in the humid tropics are not inherently more susceptible to erosion than soils of other climatic zones. Others, however, are convinced that the basic risk of erosion is greater in tropical soils. Moormann and Van Wambeke (1978) have pointed out, for example, that soil erosion is a major hazard in the tropics and that Alfisols (USDA Soil Taxonomy) are particularly vulnerable. Klingebiel (1961) has compared erosion potentials worldwide and found that under high rainfall conditions Alfisols and Ultisols (USDA Soil Taxonomy), which are widespread in the tropics, have higher erosion potentials than other soils. Recently, von Uexkull (1984) reported that Acrisols (FAO 1974 - closely equivalent to Ultisols in the USDA Soil Taxonomy) are the most important 'problem soils' of Southeast Asia, partly because they cover 51 percent of the land surface but also because of the rapid deterioration of their physical properties on exposure and their sensitivity to erosion.

Organic matter in various forms serves to bind soil particles together and thus to stabilize soil structure against splash erosion. But organic matter oxidizes much faster under the hot moist conditions of the tropics than it does in temperate lands. Consequently, fresh and decomposing organic matter, and the reserve of plant nutrients which it holds, tend to be concentrated in a thin surface layer in tropical soils. This thin layer is easily lost by sheet erosion with disastrous effects on productivity for although the soil layers beneath, unaffected by erosion, may be deep and friable, their content of plant nutrients is likely to be very low. The decline in productivity in tropical soils caused by small losses of surface soil is likely to be more rapid, therefore, than in soils of temperate zones with nutrients more uniformly distributed in depth (Lal 1985).

Whether soils of the tropics are more easily eroded than those of other regions requires further scientific study. What seems certain is that many of them are more vulnerable to damage in terms of productivity.

ii. Erosion

In the absence of adequate conservation safeguards, traditional methods of cultivation on steep slopes under heavy and intense rain usually lead to very serious soil erosion. Plots designed to measure soil loss and runoff in countries such as Jamaica, El Salvador and Taiwan have shown that traditional cultivation can result in soil losses of 100 to 200 metric tons of soil per hectare per year (Table 3). These rates of soil loss are equivalent to a loss in soil depth of some 10 mm a year, or 50 mm (2 inches) every 5 years. Higher rates of loss are to be expected on steeper slopes in the range of 25-30 degrees (47-58 percent) which it is not uncommon for small farmers to cultivate.

Small farmers are not likely to be worried about the damage which soil eroded from their fields may cause away from their farms in downstream areas. They are worried, however, by loss of production - and thus of income - resulting from soil erosion and degradation. The decrease in soil depth over time will affect their income either by reducing yields or by increasing the need for inputs (e.g. more fertilizers) to maintain production. Decrease in soil depth is readily visible to small farmers who work the soil themselves but, without advice, they may not appreciate the relationship with productivity.

Figure 1 shows in simple diagrammatic form the general relationship between soil depth, erosion and productivity. The upper diagram illustrates the steady decrease in soil depth over the years from its original uncultivated state (Do). The lower diagram reflects the fall of production from Pm (mean level of production on soils typical of upland sites). Note that production ceases (Pz=0) whilst some soil depth remains in year Yz.

Table 3 SOIL LOSS FROM TRADITIONAL CULTIVATION ON MODERATE SLOPES (from Sheng 1982)

Countries	Crops and cultivation	Soil texture	Slope	Annual soil loss tonnes/ha	Remarks
Jamaica (Smithfield) annual rainfall: 3 300 mm	Yams planted on small mounds and with clean cultivation	clay loam	17° (30%)	133	1969-1973 average
	Banana planted with clean cultivation	clay loam	17° (30%)	183	1973-1974 average
El Salvador (Metapan) annual rainfall: 1 900 mm	Maize planted with up-and-down rows and clean cultivation	loam to clay loam	17° (30%)	127	1975 Interplanted with beans in late August
Taiwan annual rainfall: 2 500 mm					
(Chi-Chi)	Banana planted with clean cultivation	clay loam	16° (28%)	92	1965-1968 average
(Fengshan)	Pineapple up-and-down cultivation	clay loam	11° (20%)	62	1957-1959 average
	Citrus orchard with clean cultivation	clay loam	16° (28%)	156	1971-1974 average
(Yuchi)	Cassava with clean cultivation	clay loam	27° (52%)	128	1963-1965 average
(Chung-pu)	Sweet potatoes with traditional furrows	sandy loam	12° (22%)	172	1951 and 1956-1957 average
(Hsinhwa)	Continuous cultivation in 2 years: sorghum, groundnuts, sweet potatoes, soybeans and maize. All clean cultivations.	loam	18° (32%)	208	1975-1976 average

Small farmers are usually too poor and too insecure of continued access to their land, to take a long-term view of their problems. Too often they feel they must maximize their immediate returns from the land, regardless of erosion. In fact, if erosion is very severe, there may be little or nothing they can do about it without outside help.

iii. Runoff

Soil erosion, unless caused by wind, is inseparable from runoff. For three reasons excessive runoff may be unavoidable from cultivated fields in the humid tropics at the height of the rainy season. Firstly, the high intensity of the rains often exceeds the normal infiltration rate of the soils. Elsewhere an infiltration rate of 25 mm (1 inch) per hour, for example, would be regarded as excellent in a soil but in the tropics, as discussed previously, rainfall intensities often exceed this. A second reason is that soils in the humid tropics are saturated or nearly saturated throughout the rainy season. Any additional rainwater can only run off. The third reason is that runoff occurs so quickly on steep slopes that soils have limited opportunity to retain and absorb the excess water. The author has observed about 25 rainy days in each of the summer months in both northwest Jamaica and northern Thailand and in many places runoff occurred after only about 5 minutes of heavy rain.

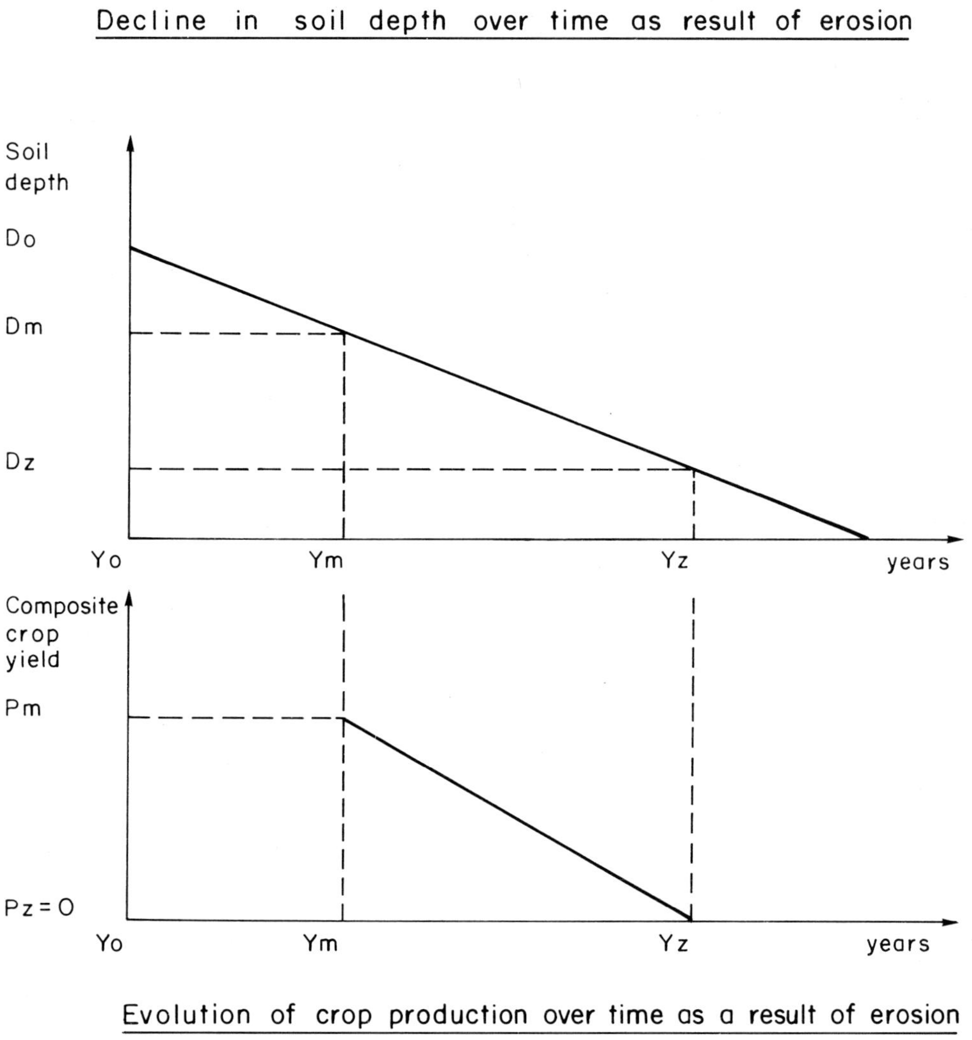

Fig. 1 Relationship of soil depth, erosion and production

The runoff coefficient, or 'C' factor, is used in the estimation of peak flows from slopes. Under conditions in the continental USA this coefficient is given the value of 0.72 for hilly cultivated land of 10 to 30 percent slope. This means that 72 percent of rainfall would be expected to run off. However, values of 0.85 and 0.90 are used in Jamaica and Taiwan for estimating peak flow from steep cultivated slopes. Tautscher (1975) has even suggested that a value of 1.0 (i.e. 100 percent runoff) should be used for designing erosion or torrent control structures in the Philippines.

Clearly such large quantities of fast and frequent runoff will rapidly erode cultivated slopes in the humid tropics unless effective conservation measures are taken to drain the inevitable runoff safely down the slopes.

2.1.4 Vegetation

The quick growth of vegetation in the humid tropics should greatly assist erosion control. However, the protective advantages of vegetative cover offered by cover cropping, green manuring, live mulching, etc., have not been widely used by small farmers as yet. Traditional ways of cultivation often call for clean weeding, burning, thorough ploughing or hoeing, all of which expose the soils to erosion.

In the humid tropics, weeds grow so fast that without frequent weeding, by hand or chemicals, the crop plants would be overgrown by weeds and crop yields much reduced. Weeds also harbour pests. Few small farmers can afford to use herbicides. The more frequent the rains the faster the weed growth and the greater the demand for frequent weeding - which exposes and disturbs surface soil and encourages erosion. This may be one of the major reasons why cultivated land in the humid tropics is eroded faster than in other climatic zones.

Burning followed by ploughing and hoeing are especially destructive operations on slopes. This kind of land preparation is usually practised just before the rainy season. The newly loosened soil is very vulnerable to the first heavy rain showers. In Jamaica, a plot newly prepared for yams (**Dioscorea cayennensis**) on a 17 degree slope was eroded at a rate of 8 tons of soil per hectare during a single 53 mm storm in April 1980.

Humid tropical vegetation often obscures erosion scars and gullies giving a false impression that erosion would not be serious local problem. Farmers and policy-makers may be deceived by this green mask of vegetation especially during the rainy season. Only close examination by trained eyes will reveal the reality of the past erosion and its continuing threat. In contrast, in arid and semi-arid regions, erosion scars are often exposed brazenly and everyone is immediately aware of the erosion hazard.

2.2 SOCIO-ECONOMIC AND CULTURAL PROBLEMS

Many socio-economic problems beset subsistence hillside farming in the humid tropics. This bulletin is concerned, however, only with the most important of those that bear upon soil conservation.

2.2.1 Land Use Problems

With the exception of rice paddies in the uplands of Southeast Asia and well-managed pastures and tree crops in other continents, most of the cultivated hillsides in the humid tropics are suffering severely from soil erosion and loss of productivity. But with growing population numbers and rural poverty, still more of this type of land is being cleared to supply the basic needs of poor farmers. FAO has estimated that between 11 and 16 million hectares of tropical forest are destroyed every year, mostly for cultivation (FAO 1981).

Acceptance of improper, erosive forms of land use in the developing countries can be attributed to one or more of the following reasons:

i. Lack of will

Governments intent on rapid development seek fast economic returns which are attained most easily by encouraging urban, industrial and lowland development. In consequence, the long-term need for protection and development of rural areas and upland watersheds tends to be neglected. The farmers themselves may not be keen to take on more arduous, expensive conservation measures; for land degradation is a long and gradual process with effects that may not be immediately discernible, even to people who have farmed the area for generations.

ii. Lack of resources

Even in countries where all concerned have a strong desire to introduce corrective measures there is likely to be a lack of know-how, manpower, equipment, funds or other resources needed to tackle erosion problems over the vast areas affected.

iii. Lack of diagnostic criteria

Ironically, though the problems of mis-use are so serious and widespread, there is a lack of practical, scientifically established criteria to assist in classifying the sloping uplands of the tropics in terms of appropriate use. Criteria for assessing land capability that have been developed in countries with different physical and socio-economic backgrounds, such as North America, may prove unhelpful and even misleading when applied elsewhere (see 3.3.2).

iv. Lack of proven conservation techniques

Once a choice of suitable land use has been made, conservation measures that will protect the slopes in perpetuity should be applied. Unfortunately, many of the soil conservation techniques tried on humid tropical uplands have proved impractical or inadequate to cope with the heavy rains, large amounts of rapid runoff and severe erosion. Some of them were developed in semi-arid regions and others are suitable only for gentle slopes and for large farms. Reliable technical information specific to the needs of small farms on steep slopes in the wet tropics is scarce.

v. Lack of tested cropping systems

Even after the land is successfully prepared for a sensible use, the choice of a cropping system that will offer assured benefits for the land and for the farmers may remain uncertain. Land productivity must be maintained, but the farmers will be looking beyond this to increased production and income. Frequently farmers ask - what better crops and cropping systems will be possible after the land is protected? Several international institutes in the tropics have conducted research into multiple cropping, improved rotations and other new systems. The work continues but widespread application of the results is delayed, mainly because of the high expenditure in labour and in technical and extension effort that the new systems usually require.

2.2.2 Land Tenure

Some systems of land tenure can cause difficulties when land improvement is attempted. Farmers who do not own their land, or at least have assured access to it, are not likely to adopt conservation measures which will cost them time, money and effort. Such work can be implemented with ease only if the landowners and the cultivators both believe that the planned measures are in their own best interests. In developing countries, where tenant farmers abound, appropriate legislation governing land tenure and soil conservation is essential.

Vast areas of national forest and other public lands are owned by the government in most developing countries. Millions of poor, landless farmers occupy, or squat, on parts of these lands and eke out a living. However, national agencies responsible for these lands commonly resist the release of any forest land for agriculture even though some of the land would be suitable for cultivation if the correct conservation and management measures were applied. It should be remembered that many forest lands were declared 'protected' or 'reserved' many decades ago, when socio-economic conditions may have been different and modern land classification procedures had not been developed. Maintenance and, indeed, extension of forest cover is one of the soundest of environmental conservation measures but this does not mean that 'forestry' is the best land use in all areas or that existing areas of national forest should necessarily be regarded as sacrosanct. To declare that forest land which is encroached upon by squatters must be used exclusively for forestry is only to prolong the deadlock between farmer and forester, especially in situations in which population pressures allow no room for the squatters to move elsewhere.

Many nations could benefit greatly from a complete reclassification of their land assets in terms of the country's current needs using practical and scientific criteria. The task may be a huge one but the benefits are commensurately large. The task can be reduced to manageable proportions by adopting a phased approach in which the work is directed in sequence to the areas of highest priority (see 3.3.3). The knowledge gained by land classification can be used to guide the best use of nationally-owned land and to assist private landowners to conserve their assets. Only with a broadly developed view of land capability is it possible to plan which parts of the total area can be safely released for cultivation, pasture or other uses, always subject to good conservation practices. Knowledge of this kind can assist decisions on whether squatters can be allowed to continue to occupy parts of the forest, or whether they must be resettled. The total numbers requiring resettlement may be reduced. For those who must be moved the land classification information guides the selection of sites where resettlement is most likely to succeed. By these means both the damage caused by squatting and the human tragedy caused by unsuccessful resettlement can be reduced.

Land tenure systems that lead to the fragmentation of land holdings cause special problems in the planning and implementation of soil conservation work. The conservation techniques that have been designed for the large, contiguous farms of the developed world need to be greatly modified before they are suited to the small farms and fragmented holdings that typify developing countries.

2.2.3 Poverty and Illiteracy

According to a World Bank report (1983), at least twenty countries in the humid tropics fall within the 'low income' category having a gross national product (GNP) **per capita** of less than US$ 320 a year. Within these countries, as in any developing country, the small farmers are among the poorest groups. Against this background, it is clear that the small farmer in a developing country cannot afford to reduce the intensity of use of his land in order to protect soils for future generations, or to protect downstream areas. Yet, if land degradation is left unchecked these people will become even poorer. Since they constitute a large section of the population in many of these countries, the economic decline of these small farmers will affect, beyond doubt, the whole country. Poverty is a self-replicating cycle which operates from the level of the farmers, up to the region and ultimately to the nation.

The high rate of illiteracy among small farmers creates difficulties in conservation extension. In developed countries the use of educational pamphlets plus a few field visits may suffice to convince a farmer of the need to practise conservation on one thousand hectares of land. The same area in a developing country may be occupied by five hundred small farmers, and several years of endeavour through repeated interviews, meetings, demonstrations, education, etc. will be called for to get across the conservation message. Whether or not the message is eventually adopted may well depend on the kind of financial and technical assistance the government is able to give.

2.2.4 Cultural and Other Problems

Over the centuries traditional systems of farming have been developed in the humid tropics. The best-known of these is 'slash and burn' agriculture - a system under which a piece of forest is cleared, crops are grown for perhaps two or three years and then the area is allowed to revert to a 'bush fallow' to replenish the soil fertility. As with most traditional agricultural systems in the humid tropics, slash and burn agriculture was effective and did little lasting damage while population numbers were low and there was plenty of land. With growing population numbers and land shortages, the conditions have changed - land can no longer be left under fallow for extended periods and without this, slash and burn cultivation can quickly lead to severe soil erosion.

However, farmers tend to cling to the traditional practices which they know and understand and are frequently slow to accept alternative practices which they may not understand or trust. Traditional practices should therefore be closely studied and every effort made to adapt and improve them to present-day conditions rather than discarding them for completely new technologies.

Local cultural and religious beliefs need to be studied carefully and understood as these can sometimes play an important part in whether or not a farming community will accept new ideas and practices.

2.3 INSTITUTIONAL PRACTICES

2.3.1 Problems of Policy and Legislation

Not many countries have clearly defined policies on land use or on soil and water conservation. Some countries may have such policies but they may not be closely or consistently followed. Conflicts of interest exist, often to the detriment of soil conservation. In some countries, for example, priority for limited funds is given to encouraging the production of export crops to earn foreign exchange as rapidly as possible with a minimum of inputs. In this short term perspective the implementation of conservation works can seem to be a waste of time and money. Equally short-term perspectives are sometimes used in the planning of other government programmes for the leasing, allocation, or resettlement of land, for example where the political aim is to settle as many people as possible in the shortest possible time irrespective of land capability or conservation needs.

Legislation which lacks consistency or firm support may prove useless. Soil conservation legislation promulgated in some countries has never been effectively implemented. Legislation has limited value unless the resources and manpower to enforce it are available. Some countries have initiated soil conservation programmes without first promulgating legislation and there can be merit in this approach. A brief statement of policy and intent may suffice at the start of a conservation programme. Later, when experience has been accumulated and serious bottlenecks have been eliminated, enactment of appropriate legislation will really help the programme.

Many developing countries may have put much emphasis on early legislation, neglecting the need for prior field experience. With further experience this premature legislation may prove impractical. A suggested approach to effective legislation is discussed in section 3.1.1.

2.3.2 Political Instability

Political instability, frequent change in government and drastic reorganization can create a feeling of insecurity and discouragement amongst professional and other government servants and disrupt all kinds of programmes. Soil conservation programmes, because of their essentially long-term nature, are particularly sensitive to the disruption which these changes cause. Fluctuation in government support for conservation programmes is common even in the absence of major government change. This makes long-term planning impossible. The situation is alleviated in many countries by the activities of soil conservation societies and other interest groups, such as 'Friends of the Earth', who continually draw the attention of government authorities to conservation needs. Public educational campaigns may also help.

2.3.3 Institutions and Technical Personnel

In many countries the responsibilities of long-established forest departments have been expanded to include a soil conservation section once the need for this work was recognized. Other countries have separate organizations responsible for soil conservation, watershed management, or the like. What is important is not the names of these institutions, or to whom they are responsible, but rather their capability to carry out the necessary field work and their management efficiency.

Problems of organization which are commonly encountered include:

- an institutional set-up which does not measure up to the size of the task;
- isolation of conservation work from agricultural development because an adequate coordination mechanism is lacking;
- shortage of funds, vehicles and other equipment necessary to implement the work;

- weakness in planning, field supervision and extension;
- most crucially, a lack of experienced, well-trained staff to do the work.

The lack of trained staff needs further explanation. The subject of soil conservation or watershed management has not been so widely taught in developing countries as agronomy or forestry, for example. Soil conservation demands an unusual breadth of basic knowledge of civil engineering, hydrology and other related sciences that lie outside normal curricula of general agronomy or biology. Training in the subject is a slow process. Experience in many countries shows that after a college graduate has been properly trained in soil conservation, he or she still needs several years of work under experienced supervisors in the field before being fit to take up an independent post. Highly competent, experienced soil conservationists are still scarce in developing countries. Consequently many conservation and watershed projects have failed mainly because of a lack of qualified, experienced staff. Lack of incentives designed to attract and retain trained personnel in the rural areas may also contribute to the staff shortage problem.

2.3.4 Infrastructure

Shortage of roads is a common problem in the uplands and marketing institutions and transportation are usually poorly developed. Without good roads farmers have problems getting produce to market and technical or extension officers find it difficult to visit or assist farmers. In many countries the building of community roads should become an integral part of any soil conservation programme. The routing, design, construction and maintenance of roads can be of utmost importance to an overall soil conservation programme since faulty design or construction can lead to landslides and catastrophic erosion in mountainous country.

Lack of drinking water, electricity and other amenities often deters qualified officers from working and staying in the uplands. The incentives and career opportunities offered to government staff should take account of these difficulties if they cannot be otherwise overcome.

These considerations underline the desirability of an integrated approach to conservation planning.

2.3.5 Funding

Budgets for soil conservation work are usually severely constrained, especially in developing countries. Sometimes they are sufficient only to keep staff in headquarters. Lack of travel funds can make the operation of a technical service difficult, not to mention financing public conservation works or the provision of financial incentives to farmers. Without adequate financial support nothing serious in the field of soil conservation can be accomplished. Ways of seeking additional funding are suggested in section 3.4.3.

Soil conservation is field oriented. Just as a bridge, no matter how well designed, needs to be built before it can benefit society, so ideas on soil conservation benefit a country little whilst they remain only on paper.

3. APPROACHES

3.1 POLICY AND INSTITUTIONAL STRENGTHENING

3.1.1 Policy and Legislation

A concise statement outlining government policy may suffice as a first step in promoting the aim of conserving a nation's soil resources. Legislation should follow, but it is rarely desirable for countries with little experience of conservation to rush into the promulgation of acts or laws. Legislation takes time and much useful work can be done in the field without it. Legislation can be delayed until experience has been obtained and bottlenecks have been identified; but it is necessary as soon as steps are to be taken to create a responsible organization, provide better coordination, and allocate budgetary resources for soil conservation. In other words, it is needed when the time is ripe for large-scale implementation.

The policy statement on conservation can be brief. Technical details can be worked out later. Land use institutions and relevant interest groups should be involved in drafting the policy. Inevitably there will be some conflicts of interest, but thorough discussion will expose the facts and benefit policy formulation. Final decisions on policy matters should be settled at Cabinet level for the issues will certainly cut across the interests of several ministries.

Some principles for government action have been established in the World Soil Charter (FAO 1982) (see Figure 2). These can be used as a basis for policy formulation and, in time, in the formulation of legislation. Some additional suggestions on policy and legislation follow:

- although the main concern may be to conserve land resources used for agriculture, other lands should not be neglected. The nation's soil resources should be seen as one entity;

- conservation priorities should be established, since governemt resources are always limited;

- soil conservation programmes should not stop at erosion control. Their production function should be emphasized and agronomic measures and soil management practices regarded as integral parts;

- proper land use and soil conservation planning should be undertaken and applied in all government land and agricultural development schemes - including land reform, land settlement, crop expansion, reservoir and highway construction, and urbanization;

- private farmers should be encouraged to adopt soil conservation measures, and the principles of government assistance stated;

- the responsibilities of various conservation organizations should be defined and coordination mechanisms described.

3.1.2 Institutional Strengthening

Institutional strengthening is needed in most of the developing countries. The size of soil conservation organizations should relate to a country's needs. The following aspects often require strengthening:

- recruitment of professionals to fulfil the basic functions of the organization. Technical posts should not be filled with unqualified or inexperienced personnel as a political expediency;

1. Among the major resources available to man is land, comprising soil, water and associated plants and animals: the use of these resources should not cause their degradation or destruction because man's existence depends on their continued productivity.

2. Recognizing the paramount importance of land resources for the survival and welfare of people and economic independence of countries and also the rapidly increasing need for more food production, it is imperative to give high priority to promoting optimum land use, to maintaining and improving soil productivity and to conserving soil resources.

3. Soil degradation means partial or total loss of productivity from the soil, either quantitatively, qualitatively, or both, as a result of such processes as soil erosion by water or wind, salinization, waterlogging, depletion of plant nutrients, deterioration of soil structure, desertification and pollution. In addition, significant areas of soil are lost daily to non-agricultural uses. These developments are alarming in the light of the urgent need for increasing production of food, fibres and wood.

4. Soil degradation directly affects agriculture and forestry by diminishing yields and upsetting water regimes, but other sectors of the economy and the environment as a whole, including industry and commerce, are often seriously affected as well through, for example, floods or the silting up of rivers, dams and ports.

5. It is a major responsibility of governments that land-use programmes include measures toward the best possible use of the land, ensuring long-term maintenance and improvement of its productivity, and avoiding losses of productive soil. The land users themselves should be involved, thereby ensuring that all resources available are utilized in the most rational way.

6. The provision of proper incentives at farm level and a sound technical, institutional and legal framework are basic conditions to achieve good land use.

7. Assistance given to farmers and other land users should be of a practical service-oriented nature and should encourage the adoption of measures of good land husbandry.

8. Certain land-tenure structures may constitute an obstacle to the adoption of sound soil management and conservation measures on farms. Ways and means should be pursued to overcome such obstacles with respect to the rights, duties and responsibilities of land owners, tenants and land users alike, in accordance with the recommendations of the World Conference on Agrarian Reform and Rural Development (Rome, 1979).

9. Land users and the broad public should be well informed of the need and the means of improving soil productivity and conservation. Particular emphasis should be placed on education and extension programmes and training of agricultural staff at all levels.

10. In order to ensure optimum land use, it is important that a country's land resources be assessed in terms of their suitability at different levels of inputs for different types of land use, including agriculture, grazing and forestry.

11. Land having the potential for a wide range of uses should be kept in flexible forms of use so that future options for other potential uses are not denied for a long period of time or forever. The use of land for non-agricultural purposes should be organized in such a way as to avoid, as much as possible, the occupation or permanent degradation of good-quality soils.

12. Decisions about the use and management of land and its resources should favour the long-term advantage rather than the short-term expedience that may lead to exploitation, degradation and possible destruction of soil resources.

13. Land conservation measures should be included in land development at the planning stage and the costs included in development planning budgets.

Fig. 2 Principles of land use and soil conservation (from the World Soil Charter, FAO 1982)

- setting up sufficient field offices to ensure that work reaches areas where it is needed and giving incentives and career advancement to those staff who are stationed and work in the field;

- establishing training centre(s) to provide basic and continuing training to the existing staff, newly recruited staff, extension officers, farm leaders, etc.;

- improving management efficiency by establishing effective communication channels, both horizontally and vertically within the conservation organization and upward to higher authorities;

- development of systems for planning, inspection, reporting, record keeping, and evaluation to monitor quantity and quality of work achieved;

- increasing work capability by supplying field stations with sufficient vehicles and equipment; together with adequate per diem and other travelling allowances for staff working in the field;

- developing close coordination amongst related organizations since many conservation works demand joint effort.

3.2 NATIONAL CONSERVATION AWARENESS

Without conservation awareness, a national programme cannot materialize. This involves people from all levels. Politicians can be very helpful if they are convinced of the necessity of soil and water conservation - and there is no lack of examples, from many countries, of politicians proving to be most influential promoters. High ranking policy-makers must also be convinced of the need for conservation. Special interest groups and professional societies need to be contacted and kept well informed. Most important of all are the general public, especially the farmers. Only when farmers are convinced and deeply believe that conservation is for their own good can such work be carried out effectively.

There are many ways to promote national awareness. Public campaigns, editorials, special articles in newspapers and magazines, pamphlets and brochures, television programmes, broadcasting, meetings, etc. can be used to educate the general public through calls for self-sufficiency, resource protection and patriotism. Eventually the public can be counted upon to put pressure on local politicians and the government for action.

For the farmers in rural areas, practical demonstrations, proper extension activities and educational meetings may be more useful. It is most important in dealing with small farmers that, once they are convinced and willing to participate in the programme, the government will be ready to help them technically and, if need be, financially. Failing this, their enthusiasm will quickly die.

3.3 SURVEY AND PLANNING

3.3.1 Land Classification in General

Some knowledge of local land characteristics - the general pattern of the climate; the nature of the topography, the soils and the vegetation; and the aims and abilities of the local population - is essential to the planning of successful conservation measures. Factual knowledge must first be gathered and then interpreted in order to distinguish lands with differing characteristics, differing potential and differing conservation requirements. Just how much knowledge is needed to provide a useful classification of the land depends on the circumstances of the study and, in particular, on the detail of planning required and on the complexity of the land itself.

A survey at national level is very unlikely to provide the detailed information necessary to plan conservation works on a single farm. Nevertheless, a broad overview at regional or national level can provide very valuable, generalized knowledge about land conditions. Such knowledge makes it possible to identify the areas where development problems, including conservation problems, are most pressing - thus helping to establish development priorities. The need for more detailed survey work, which is much more expensive and time-consuming, can then be assessed and planned efficiently and economically.

It is often best, therefore, to adopt a phased approach to land classification in which rapid overviews of large areas are followed by more detailed work in priority districts.

A national survey of conservation priorities remains to be done in most developing countries. Ideally, this should form part of a more comprehensive land evaluation study - aimed at establishing the suitability of different parts of the country for different forms of land use which are, or could be, important to the nation.

The concept of 'land suitability' as described in the FAO Framework for Land Evaluation (FAO 1976b) considers not only the physical and biological aspects of a particular piece of land but also the social and economic situation of its people. Furthermore, it recognizes that the extent to which a specified area of land can be deemed 'suitable' or 'not suitable', for a particular kind of use depends not only on the proposed crop (or other end product) but also on the planned production methods. For example, whether or not a steeply sloping piece of land is suitable for the production of arable crops by small farmers might well depend on whether essential soil conservation measures were feasible under prevailing social and economic conditions. Equally, it might depend on other factors, such as the qualities of the soils, or the reliability of the rainfall or the local availability of necessary fertilizers at acceptable prices.

Thus soil conservation needs, and the land characteristics which determine these needs, are only one group of factors amongst many that determine land 'suitability'. For further discussion of the principles and techniques of land evaluation, the reader is referred to the appropriate FAO guidelines. Separate guidelines have been published for evaluating land suitability in the context of rainfed agriculture, irrigated agriculture, forestry and extensive grazing (FAO 1983, 1985, 1984c and in press, respectively).

These ideas on broadly based land evaluation do not lessen the value of land classification in a more limited context - focussed specifically on soil conservation. Far from it. So-called 'land capability' classifications, which focus on soil conservation hazards and on the constraints which these hazards place on land use in different places, have proved their worth in many different countries. 'Land capability classification' deserves closer consideration here because it is less demanding of time and personnel than full land evaluation and is often an appropriate first step in conservation planning in developing countries. The data on which competent capability classification is based can also make a useful contribution to more comprehensive land evaluation eventually needed.

3.3.2 Land Capability Classification

The best known system of land capability classification is that developed by the United States Department of Agriculture (Klingebiel and Montgomery 1961). This system, which introduces criteria for the recognition of eight classes of land of differing conservation hazard, has been very widely used and adapted. Many countries have used the system with only slight modification. Others have copied it without considering the differences in environment which might exist between their own countries and those parts of the United States for which it was primarily designed (i.e. the Great Plains). Sometimes frustration has resulted, for whilst the principles and philosophy of the USDA classification have sound general application, the criteria proposed for distinguishing classes need to be reviewed in the light of local conditions.

'Capability' in this context refers to the range and intensity of uses that are possible on a given piece of land. Flat, well-drained land, for example, is normally subject to few conservation hazards (other than wind erosion) and can be used safely in a wide range of ways, including the intensive production of arable crops. Such land is likely to be placed in Classes 1 or 2. Very steep land, on the other hand, is certain to be highy susceptible to erosion and so offers very little choice in land use. Nearly always it must be left under natural vegetation or planted with protective forestry. Such land is placed in Classes 7 or 8.

It is important to emphasize the difference in these concepts of 'capability' and 'suitability' as they are applied to land. The recommendation in a capability classification that steep land should be planted to forestry does not imply that this is the best land on which to grow trees, or indeed, that forestry might not be the best choice of land use on flat land in the valleys. Planting trees on the steep land may deserve high development priority, nevertheless, for conservation reasons. If the term 'capability' were not so widely used it might be better to speak of 'land versatility'. With respect to land use, land of Capability Class 1 is the most versatile. It may or may not be the most productive.

If meaningful comparisons of land 'capability' are to be made it is necessary to assure a standard level of land management in each assessment. In defining the criteria for class distinctions in the original USDA system a fairly sophisticated level of mechanized agriculture was assumed to be the norm, very different from the levels of expertise and input to be anticipated on small farms in the humid tropics. Recognizing a lack of criteria suited to the tropical situation, Sheng (1972) devised a 'treatment-oriented' land capability system specifically intended for use in uplands and watersheds of the humid tropics. The system has been used satisfactorily in a number of countries including Jamaica, El Salvador, Honduras and Thailand. In the view of Hudson (1983), the system offers an acceptable alternative to the USDA system for use on slopes in the humid tropics.

The 'treatment-oriented' land capability classification is simple and practical. It is designed to be understood by field assistants and farmers. It is based on the premise that if a piece of land can be treated and protected by standard conservation practices and erosion minimized, that piece of land can be safely used for intensive cultivation. 'Standard conservation practices' in this context include bench terracing, simple structures, agronomic conservation and other modest measures (see Chapter 4). On this premise, provided that the soils are deep, lands with slopes of up to 25 degrees (47 percent) can be classified as 'cultivable', whereas steeper slopes up to 30 degrees (58 percent) are classified as capable of fruit trees, tree crops or agro-forestry. This classification is close to the reality of land use by small farmers and is unlikely, therefore, to provoke conflict between the land use wishes of government and farmers. Although soil conservation measures are a prerequisite, the classification allows some flexibility of choice in that a less intensive use may well be acceptable. Cutivable land may be used for trees or tree crops if the farmer so desires. Intensities of use that exceed the assessed capability of the land should always be discouraged.

Slope and soil depth are major factors in the determination of capability class. Stoniness, wetness, severe erosion and the risk of flooding are other factors which limit capability and so are considered as classification criteria. Since slopes and soil depths can be measured and other limiting factors identified on site, field data on land classification criteria can be collected by a junior assistant with some basic training He or she will need to have the use of a hand level and a soil auger. A trained technical officer can use the criteria so gathered to classify the lands for watershed planning.

Table 4 provides a summary of criteria used in the 'treatment-oriented' land capability classification arranged by slope classes. Appendix 3 gives a more extensive description of an application of the system in a semi-detailed study in Jamaica. A step by step approach to classification and mapping procedures has been published by Sheng (1981a). These are introduced here for reference and, hopefully, to stimulate further investigation of this subject. Readers interested in further details should consult the author's paper and report (Sheng 1972 and 1981a).

Table 4 'TREATMENT-ORIENTED' LAND CAPABILITY CLASSIFICATION CRITERIA - SUMMARY FORM

Slope class	Slope[1] ° or %	Soil depth[2] cm	Land capability[3]	Major conservation treatment[4]	Tools[5]	Land use
1	0-7° 0-12%	>15	C 1	Mainly agronomic conserv. measures; simple terraces approaching 7°	Large machine or hand	Any crop
		<15	P	Grass cover		Pasture
2	7-15° 12-27%	>30	C 2	Bench terraces & simple terraces	Medium machine or hand	Any crop
		<30	P	Hillside ditches		Pasture
3	15-20° 27-36%	>45	C 3	Bench terraces & simple terraces	Hand or small machine	Any crop
		<45	P	Hillside ditches, zero grazing, etc.		Pasture
4	20-25° 36-47%	>55	C 4	Simple terraces & some benches	Hand or walking tractor	Annual & perennial crops
		<55	P	Hillside ditches, zero grazing, etc.		Pasture
5	25-30° 47-58%	>60	FT	Orchard terraces	Hand	Tree crop
		<60	F or AF	Forest cover or agro-forestry		Trees or tree crop
6	>30° >58%	All depths	F	Forest cover		Forest only

[1] Slope limit for 'Cultivable land' can be lowered according to socio-economic and physical conditions of each country.

[2] Regardless of soil depth, any land which is too stony, wet, with severe erosion, etc. which prevents normal tillage and treatment is to be classified as pasture (<25°) or forest (>25°).

[3] C : Cultivable land
P : Pasture
FT : Land for food trees, fruit trees or tree crops
F : Forest land
AF : Land for agro-forestry

[4] For various conservation treatments, see Chapter 4.

[5] 'Tools' are those used for cultivation and land treatment.

3.3.3 Organizing Survey and Planning

The desirability of a phased approach to survey and planning has been stressed already. Starting with a broad overview and then proceeding to more detailed studies not only ensures that the expensive detailed work can be concentrated in the areas where it is most urgently needed but also that the detailed investigations are carried out within an overall procedural framework. The detailed findings in different areas can then be compared meaningfully so that experience can be transferred between areas with confidence.

In many countries it is sensible to think in terms of three phases of survey and planning prior to systematic conservation work:

i. National reconnaissance

A national overview of conservation requirements is always desirable. Since funds are certain to be limited it needs to be carried out as quickly and as economically as possible, making maximum use of all existing land survey information. The latter can be supplemented and coordinated geographically by the interpretation of up-to-date aerial photography and satellite imagery. In this way slow and costly investigations on the ground can be minimized but not eliminated. Some ground checking is nearly always necessary.

The national conservation study would desirably form part of a more comprehensive evaluation of land resources for general planning purposes which will require to be up-dated periodically as circumstances, especially economic conditions, change. The physical criteria of soils and slopes which bear directly on conservation are of a more permanent nature but they too need periodic re-examination in the light of changing economic aims and technical possibilities.

Systematic land capability mapping may not be necessary for a national overview. Indeed, the extent to which changes in slope and soils would have to be generalized at national scales of mapping may make land capability mapping unsatisfactory. It is usually sufficient merely to identify major problems and problem areas. A map at 1:250 000 scale may appropriately display the findings; this should be accompanied by a short explanatory text which includes collected data not easily presented in map form. Investigations should concentrate on:

- broad distribution of land use
- extent and degrees of soil erosion
- main local causes of erosion
- comparative conservation importance of different areas (watersheds), e.g. whether especially susceptible to erosion leading to downstream damage.

It is often convenient to carry out such surveys watershed by watershed, although much depends in this respect on the data already available. They can be undertaken quite quickly. In Jamaica (area: 11 300 km^2), for example, such a national survey was carried out in approximately 3 months, in the main by a single technical officer.

ii. Regional and watershed planning

These surveys, focussed on large priority areas, provide a basis for systematic investment planning for development and conservation. They involve systematic land capability and land suitability mapping at map scales of about 1:50 000. In addition, data are collected on present land use, conservation and protection needs, socio-economic conditions, marketing and infrastructural development.

The services of a team of scientists covering a range of disciplines are needed for this work and, although maximum use should be made of remote sensing techniques, the amount of data which has to be collected and collated can be very large. Unavoidably the work takes time. In virgin country a small survey team can be expected to cover between 600 and 6 000 km^2 in a year, depending on the complexity of the terrain, ease of access and other factors. However, if reliable information exists from previous surveys, which is true of very many parts of the world today, the work may proceed much more rapidly.

The map scale and general level of detail of a regional study are rarely sufficient to be used for planning conservation on individual small farms. However, the conservation techniques likely to be needed in the area can be identified along with detailed criteria for the selection of the most appropriate techniques at different, defined kinds of sites. Armed with this knowledge, less experienced assistants can be trained and employed to carry out detailed data collection relating to individual farms. Furthermore, realistic budgets can be drawn up for conservation work in the region. Very often a start can be made on more detailed conservation planning and implementation before the regional studies are completed.

iii. Community and farm planning

Soil conservation work is likely to be needed on community and publicly-owned land as well as on private farms. Planning for the former, which may include public erosion control work such as gully control, streambank stabilization, road verge erosion control and revegetation, is usually simpler if only because the expenses are largely borne by government. Some financial contribution from the local community may be required and, whether or not this is the case, representatives of the community should be drawn into the planning process from the beginning to ensure cooperation.

Soil conservation planning of private farms requires a different, more sensitive approach. Not only will farmers have to bear a part, perhaps most, of the cost, but they may have to accept delays in production whilst structures are installed and, perhaps, permanent changes in the use of their land possibly involving a reduction in use intensity. The techniques may be new to them. Unless they are totally convinced of the needs and advantages of change, no useful planning can be accomplished.

Thus, during farm planning, in addition to obtaining detailed information on land capability and physical conservation needs, the planner must ascertain the farmer's aims, difficulties and wishes. Note should be taken of the farmers' local knowledge of the soils on their land. The farmer's replies to questioning should be recorded; fields should be jointly visited and a practical plan should be worked out mutually. In doing so the farmer should be supplied with technical and administrative information on conservation practices, government incentives, assistance schemes and so on. Final decisions should always be made by the farmer. A simple form of agreement should be drawn up for attachment to the plans and maps on completion of planning. This needs to be carefully filed for it may be the only record of agreement between farmer and government. Appendix 1 shows the form of a questionnaire used in farm planning interviews with small farmers in northern Thailand, whilst Appendix 2 is a sample farm plan and agreement developed in southeast Jamaica. Both are for reference only; they would require to be adapted for use elsewhere.

Since most farms in the tropics are small and fragmented, time can be saved by working with groups of farmers (see 3.5.4) or within sub-watersheds. Once again the use of photo maps or aerial photographs can minimize field work on land surveying and speed the planning process.

3.4 PROGRAMMING AND FUNDING

3.4.1 A Word of Caution

The surveys and planning described in the previous section provide a basis for formulating long-term national, regional or watershed programmes. Experience shows, however, that countries which have not been active in soil conservation should not embark on an ambitious programme from the outset. Patience in gaining experience is needed. 'A small success is always better than a big failure.' Programmes of this kind should be developed only as fast as trained and experienced personnel can be made available to implement them. When large numbers of small farmers are involved, the process of planning will take time and must not be rushed. It will take a long time to regain lost confidence by the farmer if the job is not done well the first time.

3.4.2 Regular Programmes

In establishing the size of a soil conservation programme, two aspects will have to be considered: the resources available to do the work and the nation's needs and priorities.

Some countries already have large, ongoing programmes, other countries have few or none. Whatever the case is, funding is likely to be a constraint and conservation programmes will have to compete with other programmes for government funds. Under these circumstances, soil conservation programmes which are presented as just ways of preventing or curing soil erosion are unlikely to receive high priority. On the other hand, if conservation can be linked with programmes which increase production, they are far more likely to be accepted by both governments and farmers. Conservation programmes should therefore be developed and presented as a way to bring about increased and sustained production - not just as a means of preventing soil loss.

3.4.3 Alternative Sources of Funding

It is worthwhile to take a close look at all national funding possibilities if funds from regular programmes and international sources prove inadequate.

Alternative sources of national funding may include the following:

- integration in development and production projects operated by other departments;
- proceeds from a cess (or tax) on earnings of export crops such as coffee, banana, tea, pineapple, citrus, etc. which, when grown in the uplands, require conservation practices to maintain productivity;
- allocation of a small percentage of the budget raised by government for development and construction projects on sloping land. Projects concerned with road and reservoir construction, mining, housing, land settlement, etc. might be included;
- fees from city dwellers, collected from urban taxes or as supplements to utility charges, in situations where municipal watershed protection is involved;
- the organization of regions into soil conservation or watershed districts. This will assist local funding in some countries by identifying local needs more clearly and providing a fund-collecting administration.

Access to these additional sources of finance may need government policy decisions and legislation and will certainly need community support. However, the first initiatives towards their acquisition need to come from soil conservation organizations.

3.5 STRATEGIES IN IMPLEMENTATION

Development of a suitable implementation strategy is the first task of a soil conservation organization after a programme or project has been accepted. Strategies for carrying out works on public lands or on large farms can be expected to differ from those on small, private farms. Small farmers, working at or close to subsistence level, have limited resources for conservation work and, understandably, are not willing to take risks. Programmes on their land will not make progress unless farmers are strongly motivated and both technically and financially assisted.

This section will introduce some strategies, particularly for working with small farmers. Related educational, technical assistance and staff needs will be dealt with later in sections 3.6, 3.7, 3.8 and 3.9 respectively.

3.5.1 Improving Small Farm Conditions **In Situ**

A tempting way to correct improper land use and soil erosion in an upland region or watershed may be to resettle the farmers concerned elsewhere. Needless resort to this solution can be both thoughless and insensitive but often happens because planners fail to understand the real problems or are over-anxious to protect the land.

Often, in densely-populated countries, lowland sites suitable for resettlement are not available. Even in countries with low population densities the cost of properly planned resettlement for hundreds and thousands of farmers is more than many governments can afford. Even if the costs can be met, there are likely to be many socio-economic and cultural problems with which the farmers may be unable to cope in moving to a new area. In many instances the settlers soon return to the hills to cultivate the lands with which they are familiar, or they become part of the rural exodus to the already crowded cities.

In general, resettlement should be considered as a last report. A better strategy nearly always is to improve farming conditions **in situ** by discovering what can be done at farm level to raise farm production and farm incomes and minimize erosion.

Local land capability classification is the first step followed by careful conservation planning of each farm. Both activities should be carried out cooperatively with the farmers. If the criteria suggested in section 3.3.4 and Table 4 are used it will not be surprising if many hillsides come to be classified as 'cultivable'. Provided the farmer will adopt the prescribed conservation practices, these lands can be safely cultivated. Experience has shown that slopes up to 25 degrees (47 percent) can normally be cultivated safely after conservation structures are installed, except where the soils are shallow. On deep soils, still steeper slopes from 25 to 30 degrees (58 percent) can be used for growing tree crops or establishing orchards after treatment with simple conservation measures - such as the construction of individual basins and orchard terraces (see 4.2.3).

Land capability classification may show that only a part of a small farm can be classified as 'cultivable'. The planning aim, as discussed in the next section (3.5.2), must be to intensify production on this piece of cultivable land to offset the loss of production on lands which the conservation plan shows must be taken out of production. Very possibly, only those farmers who have lands with poor soils or slopes in excess of 30 degrees need to be seen as candidates for resettlement or given other work, etc. Hopefully, the total number of farmers needing to be resettled will be greatly reduced by such conservation planning.

3.5.2 Partial Intensification

Let us suppose that a small farm, 1.5 hectares in size, is found to have land of different capability disposed as follows:

F (Land only suitable for forest)	0.2 ha
FT (Land for food or fruit trees)	0.5 ha
P (Land for Pasture)	0.3 ha
C2 (Cultivable, slope 7-15 degree)	0.5 ha

Let us further suppose that, in the past, two-thirds of the farm (1.0 ha) was under cultivation with crop 'Y' and gave the farmer a return of 1X. The remaining 0.5 ha used to be in fallow. A few chickens, goats and several fruit trees yielded a further return of 1X to the farmer. Thus the total farm income was 2X.

The land capability classification calls for a reduction in the area cultivated from 10 to 0.5 ha, a reduction of 50 percent. How can the farm maintain its previous level of production and income?

The answer lies in intensifying the use of the 'cultivable' land. If the government helps (see 3.7 and 3.8), the farmer may terrace his 0.5 ha of cultivable land - yielding 0.4 ha of flat terrace and 0.1 ha of terrace risers. Government sources may also provide improved planting material of crop 'Y' (or some new, more rewarding, crop) together with a loan or subsidies for fertilizers, herbicides, etc. These might double or treble the crop yield per hectare allowing the farmer and his family to obtain the same return from the smaller area (0.4 x 2.5 = 1X). Because the cultivation work is concentrated on 0.5 ha and on gentle slopes some saving on labour in the future can be expected. This will come in useful if the government also assists the family to gradually expand the numbers of their fruit trees and animals on the remaining area of their farm, which has now increased to 1 ha. The income from the non-cultivated area could easily reach 2X. Consequently the farmer's total income might well rise to 3X. Because cultivation is concentrated on terraced land, erosion will be minimized.

Figure 3 shows the hypothetical farm and illustrates the theory.

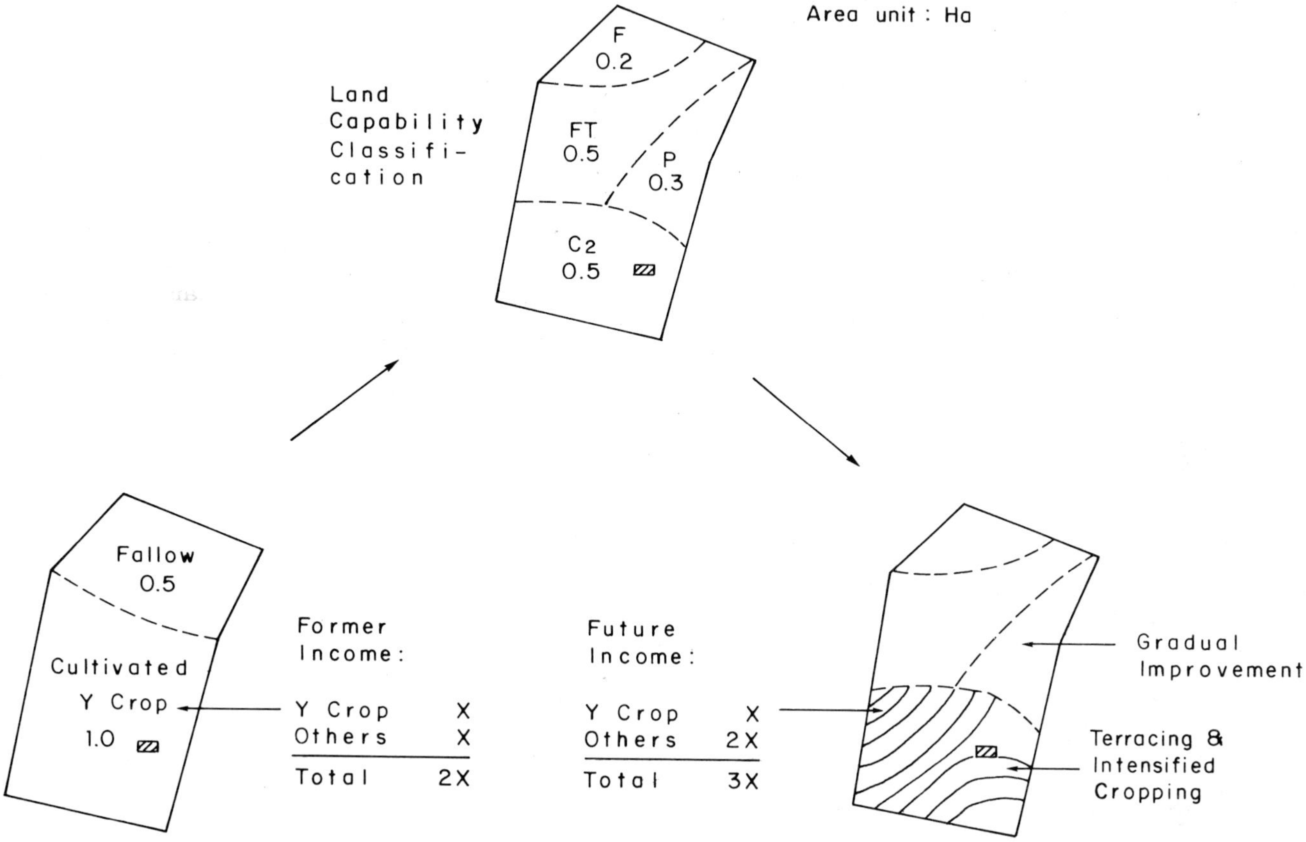

Fig. 3 Partial intensification of cultivable land in a farm

3.5.3 Progressive Treatment

Conservation treatments such as bench terracing can be labour-intensive and costly. In rural areas there may well be idle hands but the small farmer can rarely hire them to assist him. The reasons may be social but are more often economic. Furthermore, there may be seasonal labour shortages in the uplands - always when help is most needed.

Therefore, the conservation treatment of a farm needs to be carried out in a progressive manner reflecting the availability of labour. If family labour can complete the task they should be encouraged by government to do so. Experience shows that when farmers build their own terraces they are less likely to neglect future maintenance work.

Returning to the farm described in 3.5.2 (and Figure 3), as an example, the total labour requirement for the work proposed can be estimated as follows:

Bench terracing 0.5 ha of C land	225 man/days
Waterway construction	25 m/d
Orchard terracing 0.5 ha of FT land	80 m/d
Pasture improvement of 0.3 ha	50 m/d
Tree planting on 0.2 ha	15 m/d
Total	395 m/d

If the farmer's son is old enough to work on the farm and the farmer's wife is free to work half time, the available daily labour on the farm is 2.5 m/d. Food and cash crop production are a farmer's first interests and will occupy his time and mind during the main growing season. He should be encouraged, therefore, to start the construction of bench terraces and the waterway at the beginning of the dry season when farming activities normally slow down and soil moisture conditions are suitable for construction work. Working steadily the family would require 3 to 3.5 months to put in the 250 m/d required to complete the first two main items on the programme. If all goes well and the soil is managed properly, the first crop on the new terraces could be planted just as the rainy season begins. The farmer can leave the remaining conservation work to the second or third years in accordance with his interests and plans.

If the son is too young to join in the work, the farmer and his wife can build 0.25 ha of bench terraces in the first year and leave half until next year, or he can treat the whole 0.5 ha with 'intermittent terraces' (see 4.2.3) for quick protection and complete the full benches over 3 years.

Farmers should not be rushed. Unless erosion is very serious it is best controlled gradually, according to the availability of labour. Using the farmers' own labour to protect their own lands through a progressive programme of treatment should be seriously considered in all conservation programming.

3.5.4 Group Action

The involvement of groups of farmers will assist soil conservation planning, extension, supervision and implementation. Conservation works such as waterways, gully control structures and diversions frequently concern more than one farm. These will be completed more efficiently if the work on different farms is jointly planned and implemented.

Extension work and supervision will also be more effective if neighbouring farms can be organized into groups. Neighbours within a group, known to each other over a long period, will be inclined to help one another and if a minority is unenthusiastic, they will be subject to pressure from the rest. In particular, group action can alleviate peak labour shortages by encouraging alternating work periods on each others farms.

In many countries, small farmers are found to be very independent, indeed individualistic. It is often difficult to organize them into large cooperatives but it may be easier to organize them into small groups. Small groups are also easier to manage than cooperatives, especially so when a contact farmer or a farm leader can be identified within the group and given proper training and a personal incentive. Conservation work can then be organized much more efficiently.

3.5.5 Integration into Farming Systems

If soil conservation work can be integrated into existing farming systems it is likely to be accepted without resistance. Some systems already incorporate conservation principles. In Southeast Asia, for instance, level bench terraces are constructed in order to impound water to grow paddy rice. These terraces have been maintained and extended over thousands of years and are an integral part of the paddy rice farming system. Many small farmers in these areas construct and maintain such terraces on their own initiative and without government aid.

The study of farming systems as complete packages is a relatively new concept which conservationists should examine closely. If bench terracing, for example, can be introduced as part of a complete package also involving, perhaps, supplementary irrigation and new additions to the cropping system, such as new vegetables, the whole may be more acceptable to the farmers. The conservation work is then accepted along with other innovations.

Multiple cropping, close planting and the inclusion of cover crops are measures that may be worth introducing into a farming system. All generate more income and all provide a quick and dense ground cover for reducing splash erosion. Early crop varieties which form a dense cover before the heavy rains may also deserve consideration in farming system improvement.

3.6 FARMERS' PARTICIPATION AND EDUCATION

3.6.1 Farmer Participation

Farmer participation is the key to the success in any soil conservation programme. Regardless of how technically sound any plan is, it cannot be successfully implemented without the support and participation of the farmer. Any soil conservation programme must therefore contain adequate provision for such activities as discussions, meetings, interviews, training and demonstrating various practices and methods.

3.6.2 Demonstrations

"Seeing is believing". At the initial stage of a national or regional programme a number of small demonstration plots should be set up on both public and private land. On public land there should be few practical problems in implementing an ideal plan with results that will show the farmers what can be achieved and what can be confidently anticipated. However, farmers are more likely to be convinced by results from plots on their neighbours' farms. They usually trust their fellow farmers more than they do the government. Setting up the plots and implementing the demonstrations may be more difficult on private lands. The results may be less perfect but they will be seen to be closer to what the farmer can hope to achieve on his own land.

Soil conservation plots should demonstrate:

- the effectiveness of different conservation measures in controlling erosion under locally important crops and land uses;
- the benefits of conservation; in terms of improved environment, labour saved in cultivation, and increased production.

Without the latter, the farmers will not be interested. Figures for costs and returns should be carefully and clearly recorded and made known to the farmers.

Demonstration sites should be carefully selected. The sites should be representative of important agricultural zones in terms of topography, soils and land use. The locations must be in places which the farmers can conveniently visit. Once established the plots should be well maintained. Demonstration plots can be used as training grounds both for farmers and technicians. In some countries several hundred demonstration plots have been set up at the start of national soil conservation programme.

3.6.3 Education and Training of Farmers

Education and training of local farmers should start without delay as soon as an area or watershed is chosen for soil conservation or development. Such education includes village meetings, exhibitions, field visits, and the showing of slides and movies. The first objective is to motivate the farmers to participate in the programme. Activities of this kind are standard extension techniques but they are rarely applied sufficiently. Whoever is responsible for this extension work should bear in mind that soil conservation measures may require a drastic change in land use - possibly a change in the terrain itself (e.g. by terracing). It is rarely a simple matter like introducing new types of fertilizer or crop varieties. Farmers want to think twice, at least, before they take up these revolutionary ideas of conservation. Convincing the farmers is only the first step. A programme of technical and financial assistance should speedily follow (see 3.7 and 3.8), otherwise their newly-kindled enthusiasm will soon die.

Special training is usually given to selected people, such as village headmen, contact farmers, or active members of youth clubs. Selection should be done carefully. Basic requirements for contact farmers or conservation leaders are age (say, 25 to 35), physically active disposition, attitude, adequate level of education and access to a piece of land that can be used for demonstration purposes. The special training course might last between 10 days and two weeks. It should include conservation principles, techniques, and crop management. Ideally, it should be carried out in an established demonstration area. Upon completion of their training, the selected people will be expected to return to their villages to assist their fellow farmers and act as liaison people between the government and local farmers. If circumstances permit, government should provide them with a little financial support or incentive. After a year or two some of them may be given more advanced training.

A different, more generalized training course may be needed for village and community leaders. One or two days of training may suffice with the emphasis on lectures, discussions and slide or film shows but with some field visits. Participants should be the influential people in the farming community who need to be convinced of the necessity for conservation but who are not required to know the technical details.

3.6.4 Extension Services

The effectiveness of farmer education and training depends largely on the quality of the local extension officers and of the agencies which support them. Unfortunately, in developing countries, extension agencies and their officers are usually faced with many constraints and problems.

i. Extension agencies

In developing countries, extension agencies are liable to be understaffed, poorly equipped or immobilized by lack of transport. In these circumstances especially, young extension officers find it difficult to win the trust and support of farmers. The officer to farmer ratio may be as low as 1:4 000, making it impossible for the officers to meet or reach most of the farmers each year - much less work with them on their farms. For soil conservation work it is necessary to plan, design, set out and supervise the work on individual farms. This requires several visits to each farm. The situation is made worse when extension officers are charged with a dozen duties of which conservation is only one and not necessarily seen to be the most important.

Thus, to be effective in conservation, extension agencies and their working practices need to be overhauled and strengthened. Very often help needs to be sought from outside the agency, even from outside the country. In some countries, a separate soil conservation department assists extension agencies to carry out extension work and trains staff in basic conservation techniques. Some have their own conservation extension units and hire local conservation headmen to supplement the regular extension forces.

ii. Training extension officers

Since extension staff in many developing countries have received little formal training in soil conservation, short courses should be arranged for their benefit. The course work for a three week training course of this kind, which proved to be appropriate in Jamaica, is outlined for reference in Appendix 4A. After receiving such training, a junior officer should work closely with an expert soil conservation officer until he has gained enough experience in conservation extension and technical assistance to work confidently on his own.

iii. Extension activities

Extension is a continuous process. But, in individual soil conservation programmes, an extension cycle with three stages can usually be recognized. The stages, which can overlap or mingle, are as follows:

a. **Motivation stage:** using normal mass education techniques to generate farmer awareness and interest in conservation issues and to promote participation in the programme. This is primarily the responsibility of the extension agency with some help from the soil conservation department.

b. **Technical assistance stage:** actual planning, design, layout and installation of conservation measures on the farms. This is undertaken jointly by soil conservation and extension agencies in cooperation with the farmers.

c. **Follow-up stage:** assistance to small farmers by the extension agency in obtaining loans for improved seeds and other farming inputs and in marketing their produce; and by the soil conservation department in the maintenance of conservation structures and practices.

It is apparent that the extension agency has a very important role to play in the overall conservation programme. In the past, many such programmes and projects have failed because of poor extension work, either because the farmers were not motivated enough, or because lack of follow-up led to poor maintenance.

3.7 TECHNICAL ASSISTANCE

3.7.1 A Two-pronged Approach

Soil conservation programmes can be planned and implemented in two contexts. Work in the two contexts can be complementary. One approach is to concentrate effort in important watersheds or special areas. The other is to provide services nationwide, wherever interest is expressed in joining the programme. A concentrated programme normally has more impact, but a 'first come, first served approach' saves time in trying to convince reluctant farmers to join programmes. A hectare protected and made safe for production is a benefit for society as a whole, not only for the farmer immediately concerned, regardless of where it is located. Furthermore, a national programme which fails to provide national services is likely to be unfair and may be politically unacceptable. Thus, whilst special emphasis may have to be put on protecting priority watersheds and special areas, additional services should be extended to all farmers in a nation.

With these ideas in mind, soil conservation offices in many countries are located at county or district level to serve all farmers in that area. At the same time, special field offices may be established in priority watersheds, important crop zones, severely eroded areas and land development project sites to provide a concentrated degree of protection and development work.

3.7.2 Integrated and Coordinated Approach

For success, soil conservation work on farm lands needs to be linked with production and development activities to provide an overall package deal. Farmers cannot afford to be interested in conservation for itself. Neither can the governments of developing countries. Therefore, conservationists should ensure that their work is not planned in isolation towards a narrow goal of erosion control. Instead the soil conservation agency should work closely with other land use or development agencies. Conservation could become an integrated part of the work of these other agencies but it is often wise to retain a body with specific concern for conservation. Technical assistance, covering protection and production, should be given to the farmer as a package.

Technical assistance to a small farm involves many activities. Once the farmer is motivated, the task of technical assistance involves farm interviews, farm planning, conservation layout and installation, maintenance, production. Supporting services and marketing have to follow. For small farmers, successful assistance means helping them to increase their production and income. To offer conservation only aimed at protecting the soil will seldom attract a farmer's cooperation.

Experience shows that not many people at policy-making or administrative level are aware of the complexity, time sensitivity or importance of technical assistance work in this area. Close coordination between many agencies is essential. For instance, once a piece of land has been terraced and prepared, the farmer must not be kept waiting for crop loans, seedlings, fertilizer or other promised aid of this kind. Delays in aid at planting time can lead to loss of yield, or even the loss of an entire crop or of a whole season's production - which could endanger the lives of the farmer and his family. This may be obvious, but such delays may well happen in such schemes as a result of poor coordination. A small farmer who is let down in this way is unlikely to trust the outsiders in future.

Table 5 illustrates the distribution of coordinated work between soil conservation and extension agencies in Jamaica.

Table 5 EXAMPLE OF COORDINATION OR DIVISION OF LABOUR BETWEEN EXTENSION AND SOIL CONSERVATION AGENCIES FOR CONSERVATION FARMING WORK (from the Soil Conservation Division, Jamaica)

Overall planning and budgeting	Joint efforts
Farmers' meeting and extension	Joint efforts
Farmers' interviewing and farm planning	Extension officers supported by soil conservation staff
Surveying and staking	Soil conservation staff
Supervision construction	Soil conservation staff
Mapping and recording	Soil conservation staff
Soil conservation subsidy distribution	Joint efforts
Follow-up activities (i.e. cropping, credit and marketing)	Extension officers
Maintenance inspection	Soil conservation staff
Farmers' training in soil conservation	Joint efforts

3.8 INCENTIVES

3.8.1 Need for Incentives

Financial incentives in support of technical assistance are usually needed to encourage small farmers to adopt soil conservation work. These incentives can be justified on the following grounds:

- small farmers are too poor to take any risk and usually have no resources to meet additional labour or capital costs;

- many soil conservation measures involve a heavy investment in labour (e.g. bench terracing). Even the use of family labour can be a burden, and an effective loss of income, for this labour might otherwise have been invested in some other production activity or in off-farm work to gain extra money;

- soil conservation itself often yields off-site benefits and it is reasonable, therefore, to ask society in general to meet some of the costs. Certainly it would be unfair to expect the small farmers in the uplands to bear the full cost when others benefit as well;

- a farmer's income is likely to be reduced in the initial stages of conservation treatment because: (1) production is lost or delayed; (2) additional labour is needed for construction; (3) inputs, time and effort are needed to restore soil quality following soil disturbance and subsoil exposure; and (4) some actual loss in production area is likely (e.g. in terrace risers and drainageways). The early losses should be covered by later gains but this takes time and the losses should attract government compensation when they occur. Figure 4, which illustrates changes in expected income associated with soil conservation, shows the likely reduction in income in the initial period;

- financial or material incentives for small farmers can be looked upon as a system of cost-sharing between government and small farmer in conservation work.

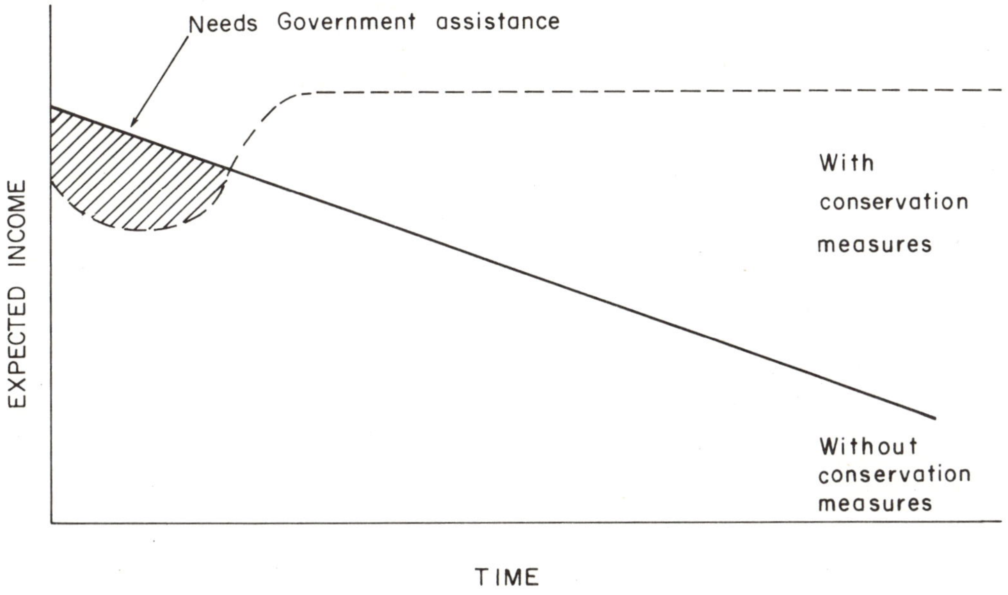

Fig. 4 Generalized curves showing expected income with or without conservation measures

3.8.2 Kinds of Incentives

Incentives may be of two kinds: direct and indirect. Among direct incentives, cash subsidies are commonly paid in recognition of work performed or as daily wages. Food, farm implements, fertilizers, etc. are given as incentives in kind, in quantities which again recognize daily labour input or amounts of work accomplished. The advantages and disadvantages of cash subsidies are discussed in the next section.

There are also many types of indirect incentives. The provision of technical assistance is one such incentive. Others include exemption or deductions from tax on income or property; farm credits; security of land tenure for farmers squatting on government land; provision of marketing services and other infrastructural development. Some of these incentives are normally built into the planning of a soil conservation programme or project.

3.8.3 Advantages and Disadvantages of Cash Subsidies

Cash subsidies are probably the most popular means of promoting soil conservation and other agricultural schemes. They are relatively easy to handle and can be distributed at any stage in the work. The main problems relate to selecting suitable criteria for payments and misuse of the cash itself. Either too much or too little subsidy can adversely affect a project, whilst misuse of funds by farmers can also affect the final outcome.

Governments in developing countries cannot afford to give large subsidies to every farmer. Experience shows that high subsidy rates tend to encourage farmers to treat more land than they can either maintain or use. Farmers may also develop a kind of 'subsidy mentality' which leads them to rely for everything upon the government. Therefore, the following questions should be answered carefully in considering the criteria for payments:

- What percentage of the total cost will be subsidized?
- What is the target for wage levels?
- Should the criteria be standardized nationwide or differ from one locality to another (e.g. important watersheds vs. other areas)?
- For how long will the subsidies be given?
- Should any limits be set per farm or per annum?

There is no universal answer to these questions. Each country needs to review its own problems in the light of available resources and determine its own criteria in particular circumstances. FAO Conservation Guide No. 12 provides a reference work on incentives (FAO 1987).

Food and other material subsidies are sometimes given. To do so is usually to ignore farmer preference, however, and can add further burdens to the administrators and field officers in the form of problems of purchasing, storage, transportation and distribution. If a food-for-work programme is already operational in the country, it will provide useful guidance and examples in the selection of criteria and in solving handling problems.

Incentives, of whatever kind, are no substitute for sound extension work and farmer education. Incentives may promote farmer participation, but the final success of a programme depends upon achieving real farmer support based on understanding.

3.9 FIELD STAFF

Much has been said about soil conservation being field oriented. The importance of the field staff cannot be overemphasized. A certain number and quality of field staff is needed to achieve any planned goal. This section will concentrate upon field staff who are concerned specifically with soil conservation although, as discussed previously, extension officers and contact farmers are also likely to be engaged in the work.

3.9.1 Work Targets and Field Staff

The number and kind of field staff needed in an area clearly depends on the volume and type of work that is planned. The majority of the work can probably be carried out by general conservationists of a subprofessional or technical level who have received some special training. Nevertheless, some professionals or specialists in agronomy, forestry or civil engineering are likely to be needed in a field office to backstop the work and to recognize and deal with problems. In El Salvador, for instance, vocational school and high-school graduates were properly trained and sent to the villages to carry out conservation work with the support of a professional team. This approach proved very effective.

Since small farmers need intensive technical assistance of the kinds listed in Table 5, the farm area which can be served by a single conservation technician is limited. As a practical guide it may be assumed that 50 ha of conservation treatment is a likely measure of the accomplishment of one field officer in a year. This measure of 50 ha could mean treating one ha on each of fifty small farms providing all were reasonably close together. It may also mean that one conservation officer with his team (e.g. contact farmers) should complete the planning, surveying and work supervision on one farm each week on average. This rough measure can be used to assess the staff requirements of a field office in relation to work targets, or **vice versa**.

Suppose, for example, that an annual target of 750 ha of various conservation activities is to be completed by a field office. The fifty hectares per officer per annum guide implies that 15 conservation technicians will be needed to complete the task on schedule. Such a group would, in fact, need to be strengthened by about 3 supervising professionals and have the assistance of about 2 other junior support staff - a total field office complement of 20. Regional and even national staff requirements can be calculated in similar manner. More realistically, perhaps, the 50 ha/officer measure can be used in similar manner to calculate the size of conservation programme that is feasible with a given number of field staff.

To establish and maintain a high level of performance in the field it is important that field workers should be fully recompensed for the actual expense and inconvenience they suffer in working away from their own homes assisting scattered groups of small farmers. Payments of per diem and other allowances should be both timely and adequate. The money involved is only a very small portion of project costs but some administrations fail to make sufficient provision for the disbursement of these allowances. This may be because they do not fully appreciate the demands of field work of this kind, but delay or failure to pay field allowances can have disastrous effects on staff morale and efficiency.

3.9.2 Training of Field Staff

The quality of the soil conservation work depends in large measure upon the quality of the technicians who plan, design and supervise it. The correct training of these technicians - the field conservation officers - is therefore of first importance.

Sections 3.6.3 and 3.6.4 have already dealt briefly with the training of extension officers, village headmen and farm leaders. In this section, the focus is on training soil conservation officers and their assistants.

Three kinds of training are usually required, as follows:

- a basic training course lasting 4 to 5 weeks is needed for conservation officers of professional and subprofessional levels. About half of this course should be devoted to field practice, one week of which may be used for field tours and visits. The usual content of such a course covers conservation principles, various survey and planning techniques, engineering structures, agronomic conservation measures, simple land use planning, soil sampling and extension techniques. Appendix 4B provides an example course which also includes watershed management subjects:

- refresher courses for the same groups of staff are also needed from time to time. Lasting one or two weeks, these courses allow revision of old subjects and introduction to new, special subjects such as the use of remotely sensed imagery, water harvesting, farm roads, etc.;

- a practical course for training field assistants or government headmen. The course should last about two weeks and 75 percent of it should be devoted to field practice and visits. By running the course at the same time as training for contact farmers (see 3.6.3) duplication of effort can be saved and the two groups can become well acquainted. A curriculum for such a course is outlined in Appendix 4C.

It is worth re-emphasizing the desirability of establishing permanent training centres within soil conservation organizations so that the conduct of training for field staff, extension officers and farmers can be continuous.

4. TECHNIQUES

4.1 BASIC CONSIDERATIONS

There are many soil conservation techniques. Some are mainly suitable for gentle slopes and large farms while others may only be useful for temperate or arid zones. For small farms on relatively steep slopes in the humid tropics the conservation techniques should take into consideration the following:

- Control of Runoff: Runoff is inevitable in the humid tropics because of frequent and intense rains. When runoff occurs on steep cultivated fields, it usually concentrates in low areas or depressions where its velocity accelerates going downhill. Rills and gullies soon develop if there are no effective control measures. Therefore, conservation techniques should emphasize the slowing down, diversion and safe disposal of runoff water.

- Integration: Although the control of runoff is vitally important, every effort should also be made to control soil detachment when slopes are cultivated. The conservation measures to be discussed, whether structural, agronomic or soil management, cannot be isolated when they are applied in the field. Soil conservation, in the broader sense, is an integrated system of land management. Also, soil conservation should be incorporated or integrated into any hill farming system for sustained production.

- Cost Effectiveness: Erosion is usually more severe on steep slopes than on gentle slopes or nearly flat land. The areas with the greatest soil erosion hazard are steep cultivated slopes under humid tropical conditions. People sometimes look for inexpensive conservation measures without considering their effect under a particular set of environmental conditions. Ineffective measures, however inexpensive, are a waste of both time and money. Therefore, cost and effectiveness should be weighed carefully at the time of planning a project.

- Opportunity for Choice: Within the limits of effectiveness, an array of soil conservation practices should be developed and provided for the small farmers' choice, according to their interest, land use plan and land capability. For instance, an older farmer who does not want intensive cultivation of food crops could choose simple conservation practices for growing semi-permanent or tree crops. On the other hand, for the farm family with plenty of labour and many children to feed, a more intensive type of conservation practices should be provided for consideration and adoption.

- Production function: Soil conservation should not stop short at soil conservation for erosion control's sake only. It should be considered as a means of achieving higher and sustained farm production, otherwise small farmers will not be interested. Therefore, conservation techniques should benefit both crop production and farm management.

4.2 STRUCTURAL MEASURES

4.2.1 Need for Structures

As mentioned earlier, in the humid tropics the precipitation rate of intense rains often exceeds the infiltration rate of the soils. During the rainy season, frequent rains saturate the soil. Excess runoff is therefore inevitable, especially on steep slopes.

If the land is covered with dense vegetation, erosion is minimized because vegetation protects the soils from splash erosion and detachment. Even when runoff occurs, the water is mostly clear. The conditions under cultivation are different. First, the soils are loosened by frequent tillage or disturbed by weeding. Secondly, the crop canopy may not provide an effective ground cover when the rainy season starts,

1. LEVEL BENCH TERRACES

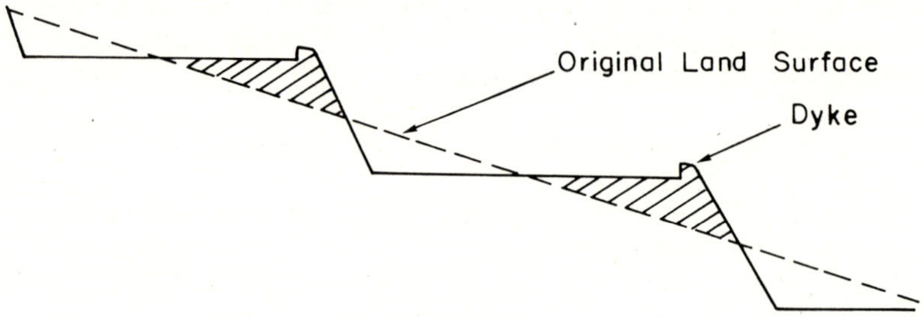

2. OUTWARD SLOPING TERRACES

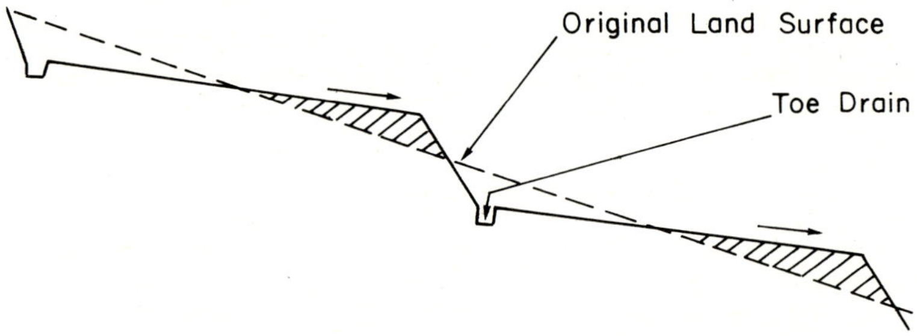

3. CONSERVATION BENCH TERRACES

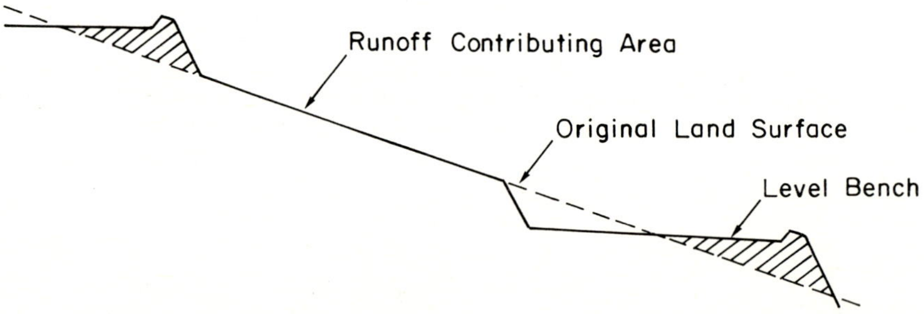

4. REVERSE SLOPING TERRACES

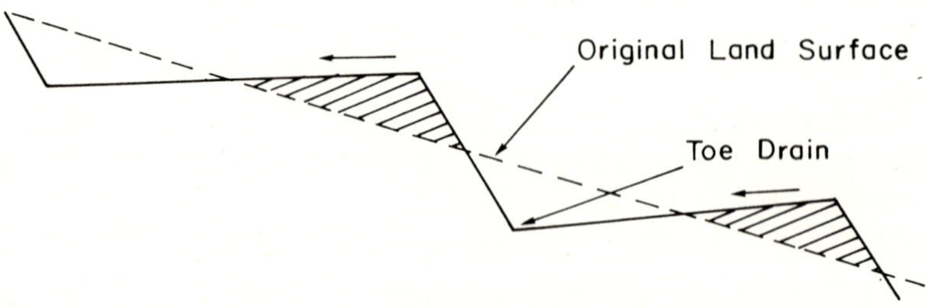

Fig. 5 Types of bench terraces

because of planting distance or time of planting. Thirdly, crops are not permanent - unlike forests and grass which grow long enough to develop a dense root system and produce ground litter which are beneficial for erosion control.

Since cultivation and runoff are both inevitable, structural measures are often needed in the humid tropics. The main functions are to divert, intercept, slow down, store temporarily, or safely dispose of the runoff thereby minimizing erosion. With some structures, such as bench terraces, erosion will not be increased significantly even when the soils are loosened by tillage.

There are two types of structures and both are needed for soil conservation purposes. The first are the land treatments e.g. terraces, hillside ditches, contour dykes all of which are built along the contour or on a grade. Their main functions are:

- to change a steep slope to a series of flat strips along the contours across a slope to facilitate cultivation and for erosion control; or

- to change a long slope to a series of shorter slopes by the use of bunds or ditches to intercept and divert runoff for safe disposal, hence minimizing erosion.

The other type of structure aims to drain runoff from the slope from where it has been concentrated. Man-made waterways and gully control structures etc. belong to this kind. Their main function is to protect and stabilize the channels for the safe disposal of runoff.

Many kinds of mechanical structures need auxiliary conservation measures such as agronomic measures and revegetation (e.g. grass planting) to enhance their erosion control functions.

4.2.2 Bench Terraces

Bench terraces are a series of level or nearly level strips running across a slope at certain vertical intervals. The level strips are used for cultivation; they are supported by steep banks or risers made of earth or rocks.

There are four main types of bench terrace, i.e. level, outward-sloped, runoff conservation and reverse-sloped types. Figure 5 shows their simple cross-sectional views. The level bench terraces have been extensively used for rice paddies, whilst the outward-sloped and reversed-sloped ones are particularly suitable for upland crops in the humid tropics.

The application, specification, construction, and cost/benefit of the reverse-sloped terraces, are described briefly in the following sections.

i. Conditions and sites for applications

Reverse-sloped bench terraces are particularly suitable for countries having the following conditions:

- severe erosion hazard due to torrential rains
- steep slopes under cultivation
- expansion of food and upland crops into hilly marginal lands
- dense populations and small holdings
- high unemployment rates in the rural areas

Suitable sites for such terracing are as follows:

- slopes between 7-25 degrees (12 percent - 47 percent). For slopes below 7 degrees the use of simpler structures is more cost effective, whilst above 25 degrees, maintenance and use of this type of terrace will be difficult;

- relatively deep soils;

- sites not too stony, wet or dissected by gullies.

ii. Specifications

The specifications of reverse-sloped bench terraces, including the grades along the terrace and reverse slopes, the width and length limits, the riser slopes etc. are shown in Table 6 together with other structural measures. A cross-section is shown in Figure 6. Below the diagram, various formulae for computing vertical interval, riser height, net area of benches and volumes to be cut and filled are shown. For quick reference, Figure 7 gives volumes and net areas at various slopes.

iii. Construction

Bench terraces can be built by manual labour, animal-drawn tools, (scraper, plough, etc.) or machines. For government-directed operations, machines are sometimes used. Small farmers mostly use manual labour and animal traction to construct terraces. Details of construction procedures together with survey and layout techniques can be seen in the following two FAO publications:

- FAO Conservation Guide 1: Guidelines for Watershed Management, pp. 154-159 (FAO 1977a);

- FAO Watershed Management Field Manual: Slope Treatment Measures and Practices* (FAO 1988).

iv. Cost and benefit

The cost of terracing per unit area is dependent upon slope, soil, width of bench, presence of rocks or tree stumps and the tools to be used. However, for a general estimate or for budgeting purposes, an average cost should be sufficient. Once the volume to be cut and filled per unit area is obtained (using the computation in Figure 6 or from Figure 7), the rough construction cost can be calculated by using the following equation:

$$C = \frac{V}{T} \times R$$

where C: Cost of cutting terraces per unit area
V: Volumes of cut and fill per unit area
T: Output per manday, animal day or per machine hour
R: Wage per manday, per animal day or rate per hour

On average, a man with hand tools can cut and fill 3 to 4 m^3 in an 8 hour period and a man with animal draft tools can complete 12 to 16 m^3 in the same period. For machine-built terraces, the output for a medium sized machine (such as a Caterpillar D-6 Bulldozer) is about 40 m^3 per hour whilst that for a smaller machine (such as John Deere 450) is abut 20 m^3. These figures can be modified according to work conditions and efficiency.

To the cost of rough cutting, several additional mandays or machine hours need to be added for shaping and riser stabilization. Sometimes top soils are worth preserving for which an additional input of 40 mandays or 14 machine hours is needed per ha for replacing topsoil.

Bench terraces are the most effective structural erosion control measures. They can reduce erosion by 90-95 percent or more (Jiang **et al.** 1980; Sheng, 1981b; and Wu 1986). Also, bench terraces are the most labour intensive; under a government cost sharing or subsidized programme, many employment opportunities can be generated for the rural population.

---

* There is a special chapter on 'Terraces and Ditches'. Detailed tables on various specifications of terraces are included for design and construction guidance.

Table 6 SPECIFICATIONS AND APPLICATIONS OF VARIOUS TYPES OF LAND TREATMENT STRUCTURE

Kind	Specifications						Applications	
	Width of flat bench	Length	Grade along terrace	Reverse slope	Riser slope	Land slope	VI[1] or spacing	Auxiliary treatments
Bench terraces								
(a) Hand made	2.5-5.0 m	<100 m	up to 1%	5%	0.75:1	7°-25°	$\frac{S \times W_b}{100-S \times 0.75}$ [2]	-
(b) Machine built	3.5-8.0 m	<100 m	1%	5%	1:1	7°-20°	$\frac{S \times W_h}{100-S \times 1}$	
Hillside ditches	1.8-2.0 m	<100 m	1%	10%	0.75:1	<25°	$\frac{S+4}{10}$ or $\frac{S+6}{10}$	Agronomic conservation measures[3]
Individual basins	1.5 m (round)	-	-	10%	0.75:1	<30°	Distance between tree rows	Hillside ditches. Orchard terraces and agronomic conservation measures
Orchard terraces	1.75 m	<100 m	1%	10%	0.75:1	25°-30°	11-13 m along slope	Agronomic measures. Individual basins
Intermittent terraces	2.5-5.0 m	<100 m	1%	5%	0.75:1	7°-25°	3 times BT	Agronomic measures. Individual basins
Convertible terraces	3.5 m	<100 m	1%	5%	0.75:1	7°-20°	as hillside ditches	Individual basins. Agronomic measures
Natural terraces	8-20 m	-	-	-	0.75:1	<7°	1 m VI	Agronomic measures

[1] VI is vertical interval between two succeeding terraces, which determines spacing.

[2] S : slope as percentage; W_b : width of bench.

[3] To be applied mostly between terraces (or on the individual basins) such as contour planting, close planting, cover cropping, mulching, etc.

CROSS-SECTIONAL VIEW OF BENCH TERRACES

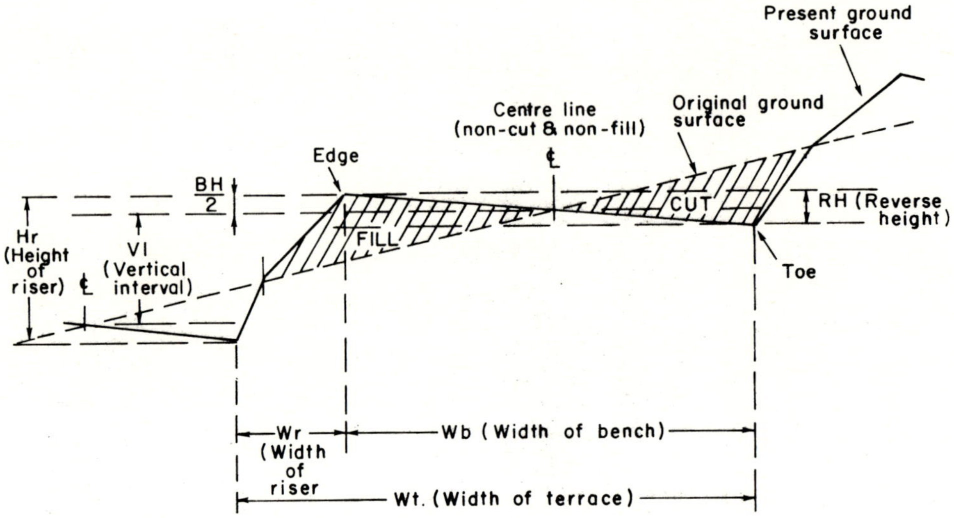

Length Units in metres except where stated

1. Vertical Interval (VI) : $VI = \frac{S \times Wb}{100 - S \times U}$ (S : Slope in %, U : 1 or 0.75)

2. Reverse Height (RH) : $RH = Wb \times 0.05$

3. Height of Riser (Hr) : $Hr = VI + RH$

4. Width of Riser (Wr) : $Wr = Hr \times U$

5. Width of Terrace (Wt) : $Wt = Wr + Wb$

6. Linear Length (L) : $L = \frac{10\ 000}{Wt}$ (per ha)

7. Net Area of Benches (A) : $A = L \times Wb$

8. Percent of Benches (Pb) : $PB(\%) = \frac{A}{10\ 000} \times 100$ (per ha)

9. Cross-section of Terrace (C) : $C = \frac{Wb \times Hr}{8}$

10. Volume to be cut and filled (V) : $V = L \times C$

Fig. 6 Cross-sectional view and computations of bench terraces (reverse-slope)

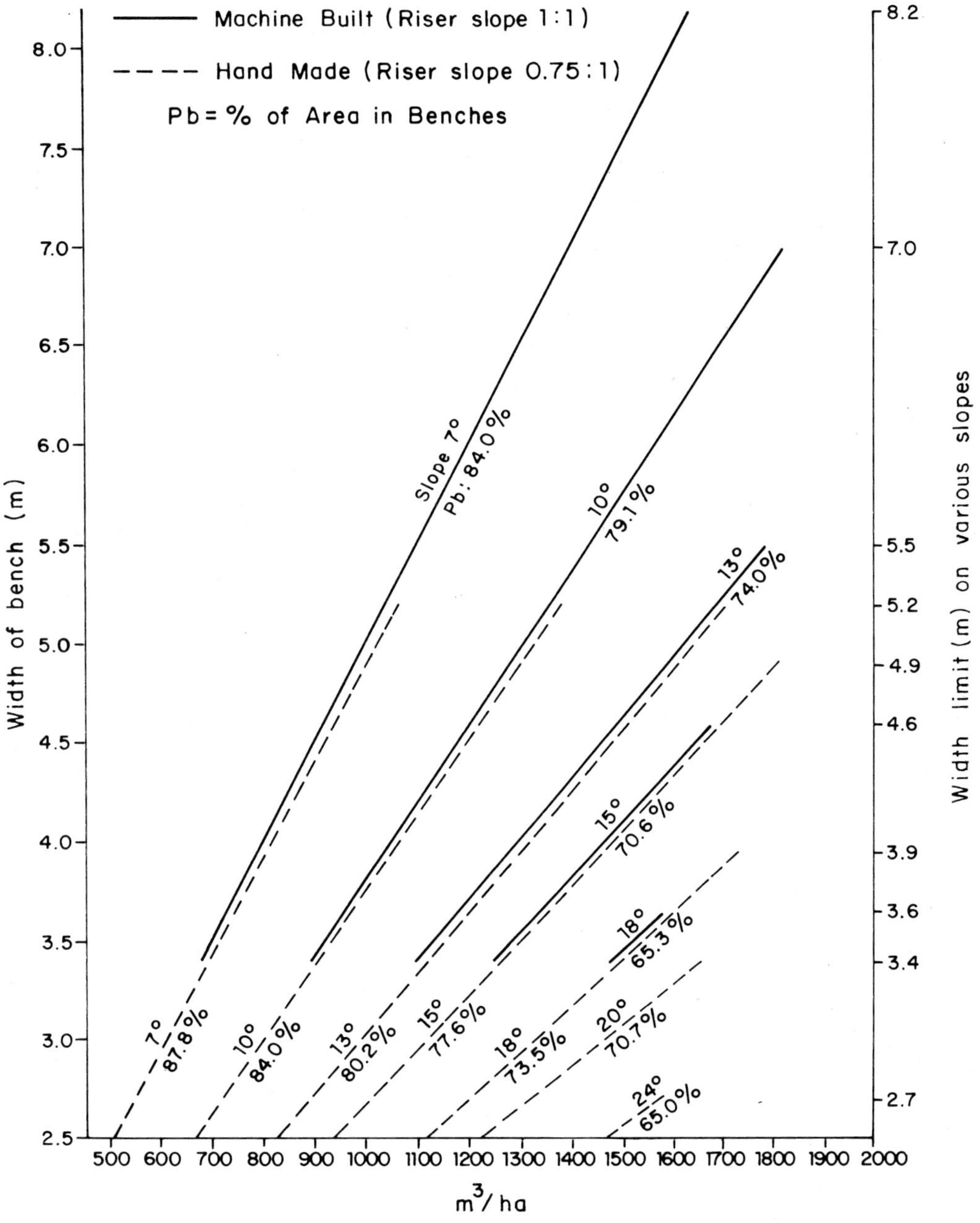

Fig. 7 Volumes of soil to be cut and filled per hectare for bench terraces

In addition to the above, bench terraces are unique among other structural measures because of the following benefits:

- creating cultivable lands from hills and marginal slopes;
- increasing crop diversity or providing free choice of crops once sloping land becomes flat;
- retaining soil moisture and fertilizers hence greatly increasing crop production;
- enhancing irrigation and mechanization potential of the land;
- encouraging settled farming and reducing the damage caused by shifting cultivation and fire;
- allowing for intensified land use without causing serious erosion.

Some of the above mentioned benefits can be quantified for economic analysis against the cost. Bench terraces with proper soil management have increased crop production by 200 to 300 percent in Jamaica. The terracing costs can normally be paid off in a short period, and the internal rate of return of bench terraces ranges from 12 percent to 22 percent according to slope (de Graaff 1981).

4.2.3 Simple Terraces and Structures

Bench terraces have many advantages but they can be costly. The initial investment is heavy for farmers and for governments. There is always an interest, therefore, in relatively inexpensive, simple conservation measures that farmers and governments feel they can afford. Whether the simple measures are efficient is another matter. To employ ineffective measures, though inexpensive in themselves, can lead to a considerable waste of resources. Vegetative barriers, strip cropping and contour cultivation, for example, have often been shown to be ineffective in cultivation on steep slopes under humid tropical conditions (Champion 1966; Liao 1976; Wu 1986).

By supplementing simple structures with agronomic measures (see 4.3), however, the aim of considerably reduced costs together with a high level of erosion control may be achieved. Controlling erosion by partially controlling both soil detachment and transportation can be most cost effective. For example, the average cost of hillside ditches is only one fifth that of a bench terrace on the same slope, yet research has shown that they can reduce erosion by 80 percent, or more, when used together with agronomic conservation measures (Sheng and Michaelsen 1973; Liao 1976; Wu 1986).

The descriptions and diagrams which follow are of six simple terraces or structures which can be used by small farmers on various slopes. Some are discontinuous types of terrace and others are transitional. Their specifications are given in Table 6 and their uses are summarized in Table 7. As mentioned earlier, it is essentially up to the farmer to choose which of the structural measures should be used, taking into account his cropping interests and other needs.

i. Hillside ditches

Discontinuous type of narrow, reverse-sloped bench terraces built across the slope in order to break a long slope into shorter ones. The narrow benches are mainly for intercepting and diverting runoff. They are better than the conventional trench type which cause maintenance difficulties and work inconvenience. Being a bench type of ditch, they can be used for footpaths or push-cart roads and occasionally for cultivation. The distance between two ditches is determined by the slope. Crops may be grown between hillside ditches if supplemented with agronomic conservation measures. This land treatment can be applied on slopes of up to 25° (47 percent). The calculation of distances between

Table 7 SIMPLE TERRACES AND STRUCTURES AND THEIR MAJOR USES

Treatment	Major Use
Hillside Ditches	For upland crops, particularly semi-permanent crops up to 25° (47 percent) slope. Usually they are applied in conjunction with agronomic conservation measures. They can also be used for protecting new pastures.
Orchard Terraces (Modified bench terraces)	For fruit trees, food trees or tree crops on steep slopes, 25° to 30° (47 to 58 percent).
Individual Basins (Crescent/eyebrow terraces)	For tree crops and other large plants. They can be applied on shallow soils and in dissected terrain. Usually they are supported by hillside ditches, orchard terraces, etc. on slopes up to 30°(58 percent).
Convertible Terraces	For mixed farming or for flexible land use, from 7° to 20° (12 percent to 36 percent) slope.
Intermittent Terraces	For upland crops or partly irrigated crops where the structures are to be built over several years, from 7° to 25° (12 percent to 47 percent) slope.
Natural Terraces (Contour dykes or earth/stone bands)	For upland crops or semi-permanent crops on gentle slopes, below 7° (12 percent). They form terraces gradually over several years of cultivation. Occasionally, they can be applied on slopes more than 7°, but they should be used with extreme caution.

ditches on various slopes, cross sections and volume, etc. are shown in Figure 8; sketch plans and the layout of hillside ditches can be seen in Figure 9.

ii. Orchard terraces

Orchard terraces are also a discontinuous type of narrow reverse-sloped bench terraces which are applicable on slopes up to 30° (58 percent). The space between two terraces is determined by the planting distance of the tree crops or for the convenience of the crop management. As they are generally used on steep slopes, the spaces between them should normally be protected by a permanent vegetative cover such as grass, legumes, or other cover crops. The main tree crops are preferably planted in individual basins (see next section). One orchard terrace will serve two rows of trees (see Figure 10). The calculations for the specifications are shown in Table 8.

iii. Individual Basins

Individual basins (sometimes called 'crescent' or 'eyebrow terraces') are small round benches for planting individual plants. The diameter of the basin is adjusted to the needs of the crop. They are particularly useful for establishing semi-permanent or permanent tree crops on steep slopes for erosion control. The basins retain soil moisture, particularly if they are mulched, and reduce the need for weeding. They also prevent fertilizer being washed away. They can be

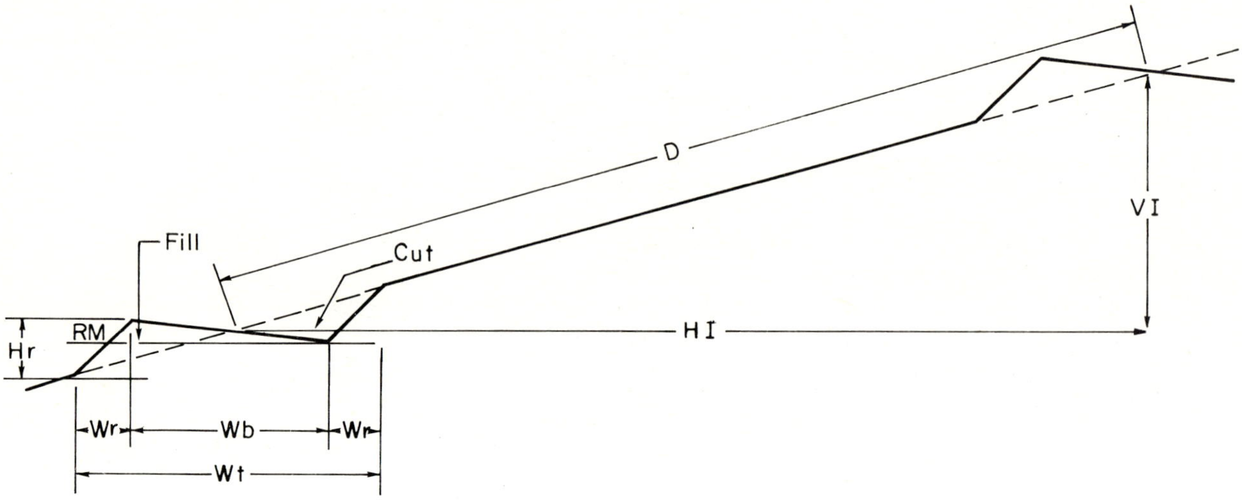

SYMBOLS AND COMPUTATIONS

Width of Platform	:	Wb	=	2 m
Theoretical Vertical Interval	:	TVI	=	$\frac{S \times Wb}{100 - S \times U}$ (S: slope in %; U: 0.75 or 1)
Reverse Height	:	RH	=	Wb x 0.10
Height of Riser	:	Hr	=	(TVI + RH)/2
Width of Riser	:	Wr	=	Hr x U
Width of Terrace or Ditch	:	Wt	=	Wb + 2 Wr
Vertical Interval between two Ditches	:	VI	=	(Sx4)/10,(S+6)/10, etc.
Horizontal Interval	:	HI	=	(VI/S)x 100 (or VI/tan of slope angle)
Inclined Distance	:	D	=	VI/sin of slope angle
Linear Length	:	L	=	10,000/HI (per ha) or 1,000/HI+Wt
Area of Platform	:	A	=	L x Wb
Percent of Platform	:	Pb	=	(A/10,000) x 100 (per ha)
Cross Section of Ditch	:	C	=	Wb x 2 Hr/8
Volume of Cut and Fill	:	V	=	L x C (per ha)

Fig. 8 Cross-section and computations for hillside ditches

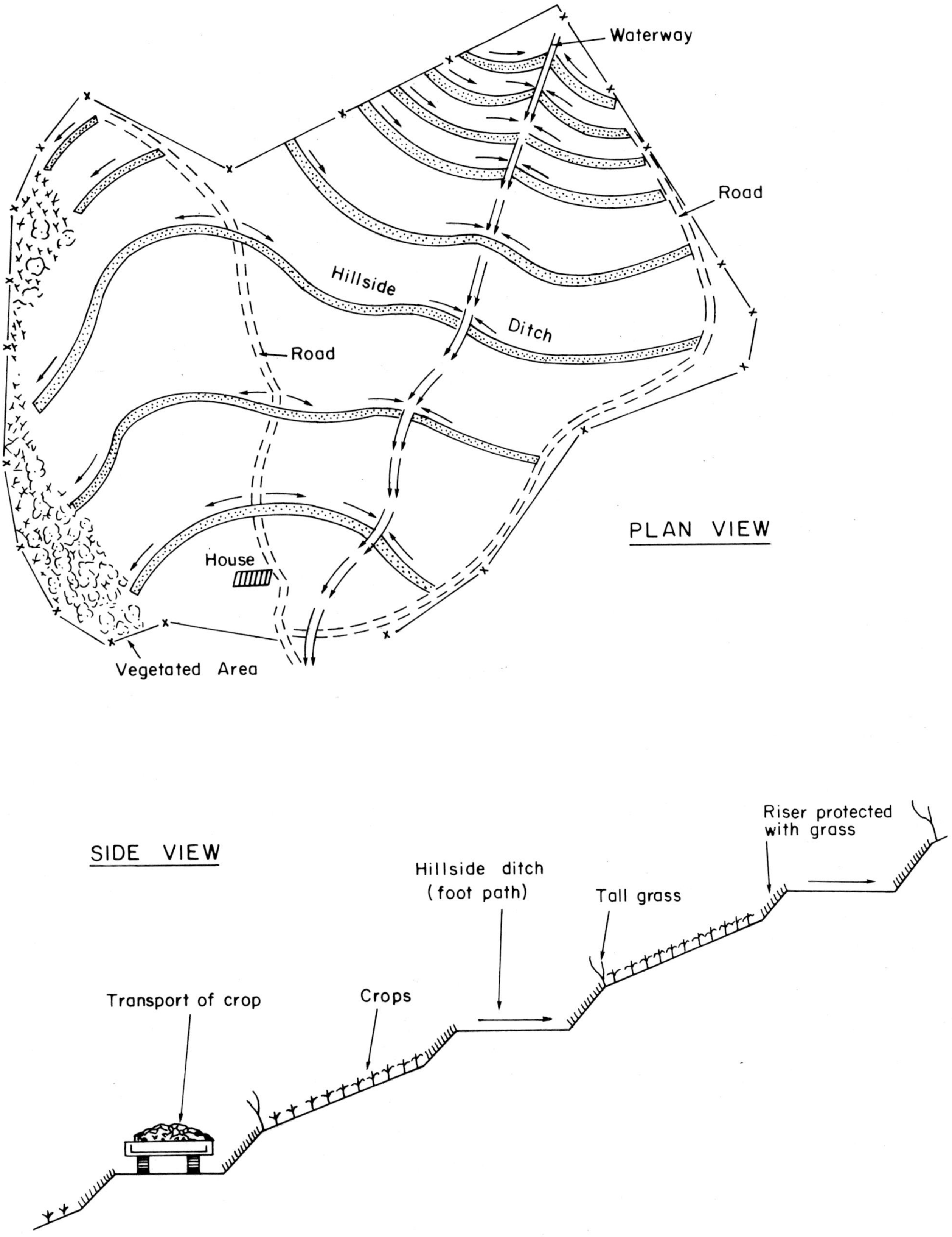

Fig. 9 Plan and layout of hillside ditches

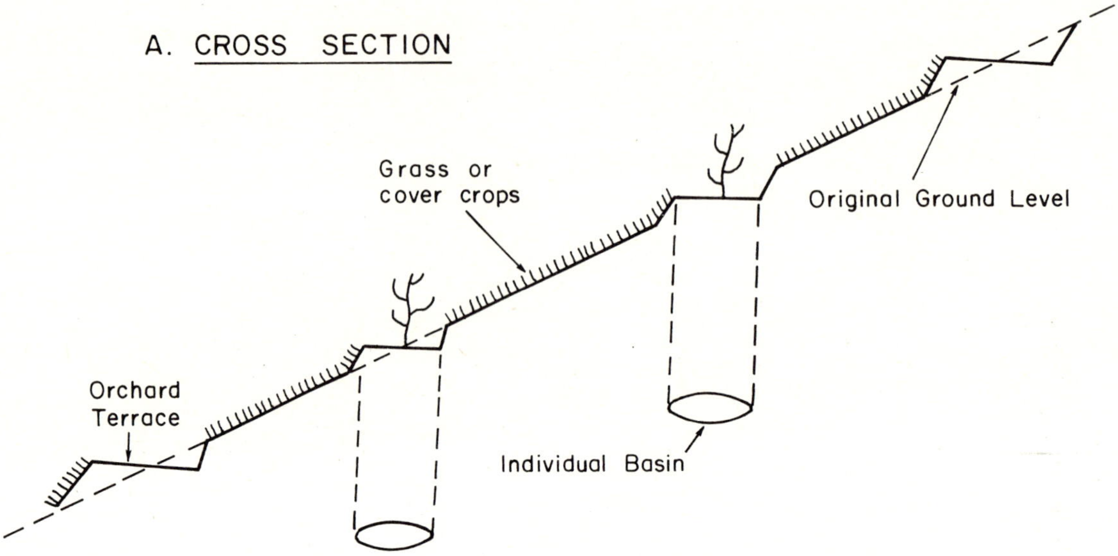

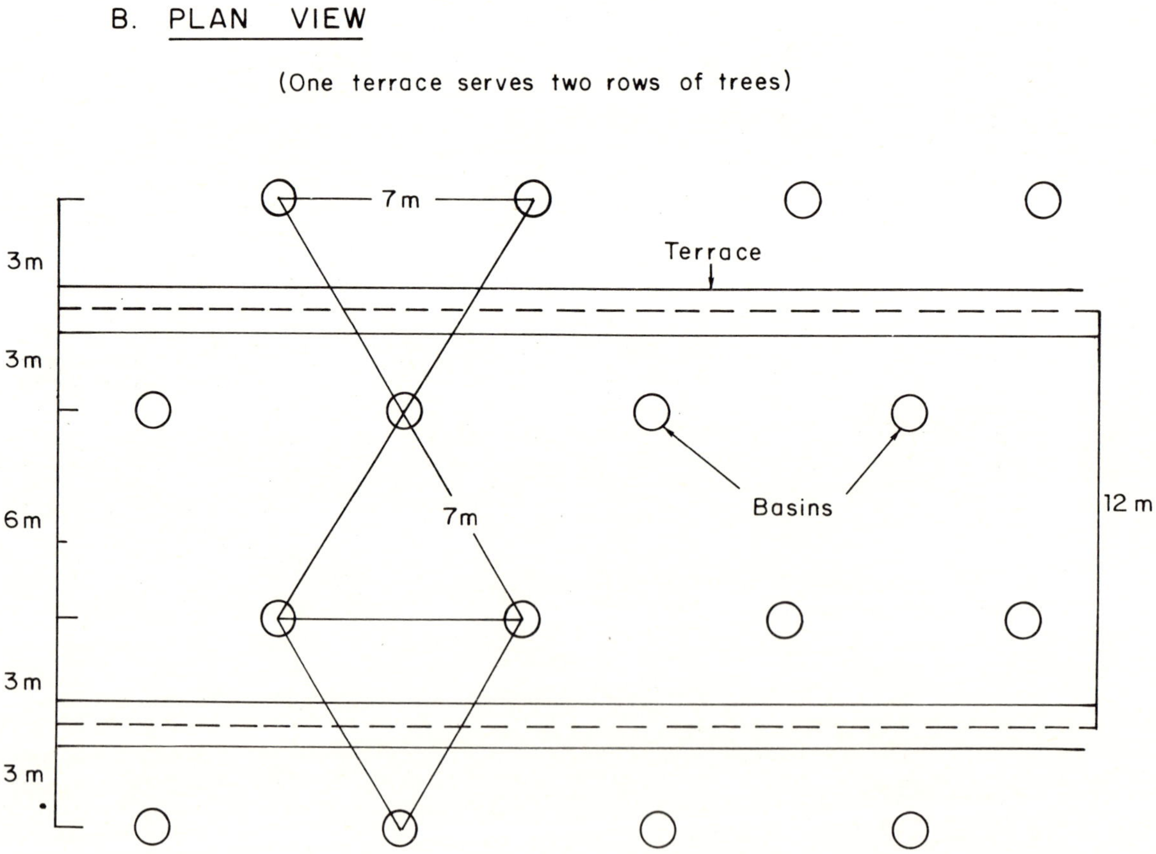

Fig. 10 Cross-section and plan views of orchard terraces

Table 8 COMPUTATIONS OF ORCHARD TERRACES

Width of Platform	:	Wb	=	1.75 m
Theoretical Vertical Interval	:	TVI	=	$\frac{S \times Wb}{100 - S \times U}$ (S: Slope in percent; U: 0.75)
Reverse Height	:	RH	=	Wb x 0.10
Height of Riser	:	Hr	=	(TVI + RH)/2
Width of Riser	:	Wr	=	Hr x U
Width of Terrace	:	Wt	=	Wb + 2 Wr
Horizontal Interval	:	HI	=	10.5m, 12m, or 14m
Linear Length	:	L	=	10,000/HI (per ha)
Area of Platform	:	A	=	L x Wb
Percent of Platform	:	Pb	=	(A/10,000) x 100
Cross-section of Terraces	:	C	=	Wb x 2 Hr/8
Volume of Cut and Fill	:	V	=	L x C (per ha)

constructed on relatively uneven or dissected terrain and on shallow soils. Normally, they would be supported by hillside ditches or orchard terraces to control excess runoff, and also by cover crops between the basins. Figure 11 shows a diagram of an individual basin together with volumes and the estimated output per manday. A photograph of a combination of individual basins with hillside ditches is shown in Plate 3.

iv. Convertible Terraces

Convertible terraces are widely spaced benches alternating with strips of untreated hillside for mixed farming. The distance between two benches is calculated in the same way as hillside ditches. The benches are used for growing staple food whilst the unchanged slopes between are used for semi-permanent crops or tree crops. In the future, should the farmer wish to cultivate more food crops, he or she can convert the remaining slope to bench terraces. On the other hand, the farmer may place all of the area under tree crops should age, labour, or economics present problems for intensive farming. This transitional type of terrace allows a choice between intensive or extensive use to be made in the future. Figure 12 shows a cross-sectional view. Volumes per ha can be calculated in the same way as for hillside ditches but using a 3.5 m wide bench.

v. Intermittent Terraces

Fundamentally these are bench terraces but with construction spaced out over a period of three or four years. Out of every three terraces staked out, only the top one is constructed, leaving the two below for future construction. Both the spaces between and the benches can be planted, but with different types of crops The benches will intercept runoff from the slopes above for erosion control and also for water conservation if required. Plate 4 is a photograph of intermittent terraces with taro (**Colocasia antiquorum**) planted on the slopes and upland rice growing on the benches. A cross-sectional view can be seen in Figure 13. Specifications for intermittent terraces can be calculated using the basic bench terrace equations.

vi. Natural Terraces

Natural terraces arise when low dykes (bunds) or dense grass lines (e.g. of Vetiver) are constructed along the contour or on gentle slopes using either earth or locally available rocks. They are designed and constructed in such a way that the top of the lower dyke is level with the middle of the slope to the next dyke above (see Figure 14). A terrace will form by natural infilling after several

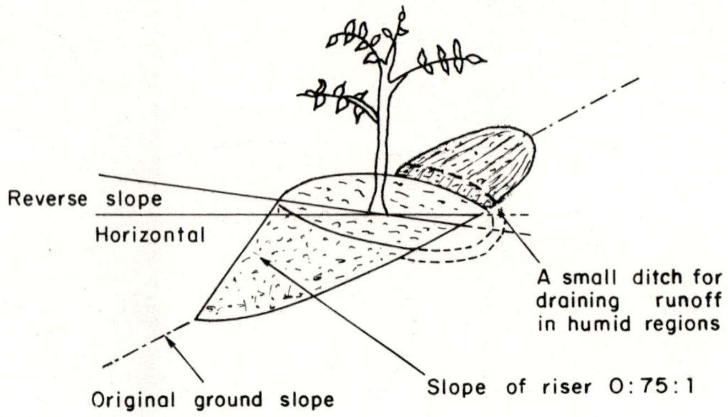

Degree of slope (°)	0.9 m (3ft) Dia. Vol.	No/day	1.5 m (5ft) Dia. Vol.	No/day	2.1 m (7ft) Dia. Vol.	No/day
	m^3		m^3		m^3	
10	0.024	112	0.112	24	0.307	9
12	0.028	96	0.132	20	0.361	7
14	0.033	82	0.146	18	0.420	6
16	0.038	71	0.177	15	0.484	6
18	0.044	61	0.202	13	0.554	5
20	0.050	54	0.230	12	0.630	4
22	0.056	48	0.261	10	0.715	4
24	0.064	42	0.295	9	0.810	3
26	0.072	38	0.334	8	0.917	3
28	0.082	33	0.379	7	1.041	3
30	0.093	29	0.431	6	1.184	2

Remarks:
1. Riser slope: 0.75:1
2. No/day: No. of basins per manday
3. Average output: using 2.7 m^3 per manday

Fig. 11 Diagram of individual basin and volumes of construction

Plate 3 Individual basins planted with young bananas and supported by hillside ditches (Jamaica)

Plate 4 Intermittent terraces are planted with taro on original slopes and upland rice on benches (Northern Thailand)

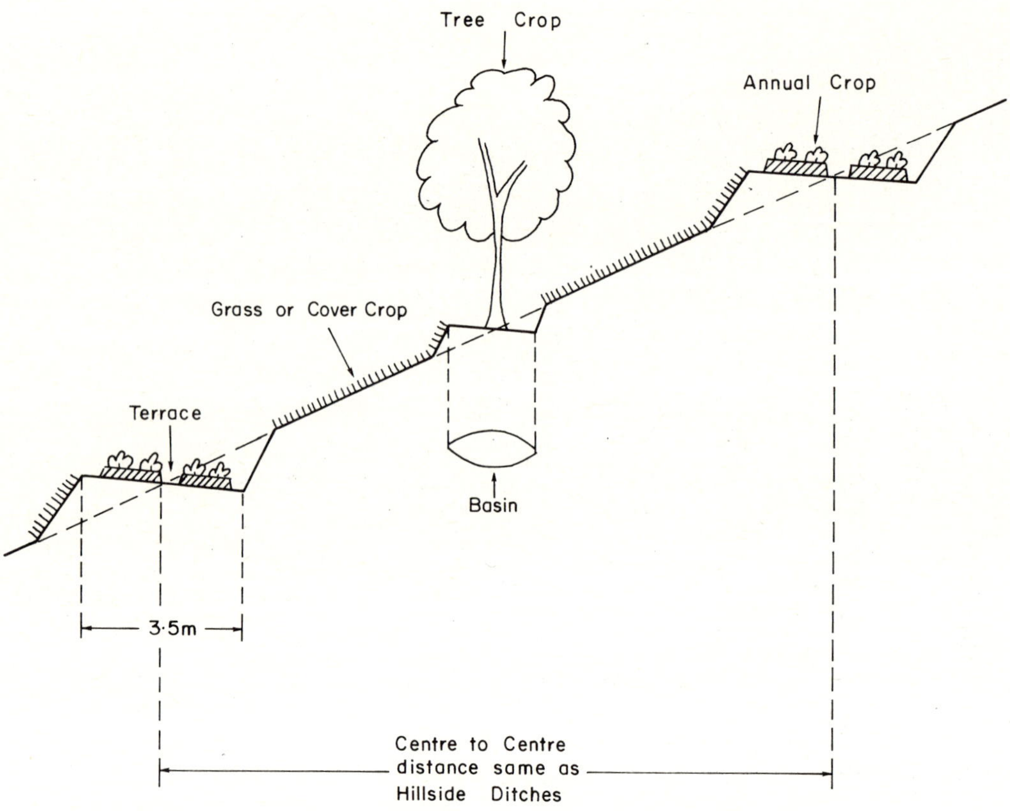

Fig. 12 Cross-sectional view of convertible terraces

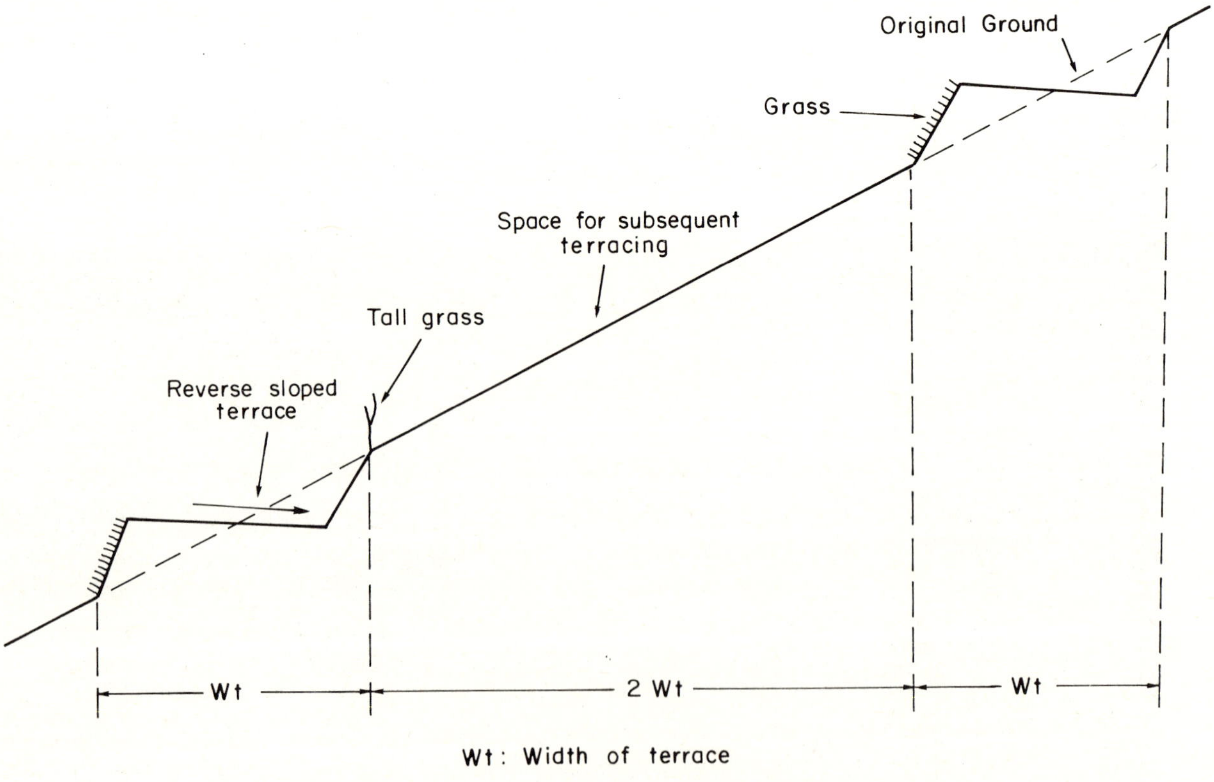

Fig. 13 Cross-section of intermittent terraces

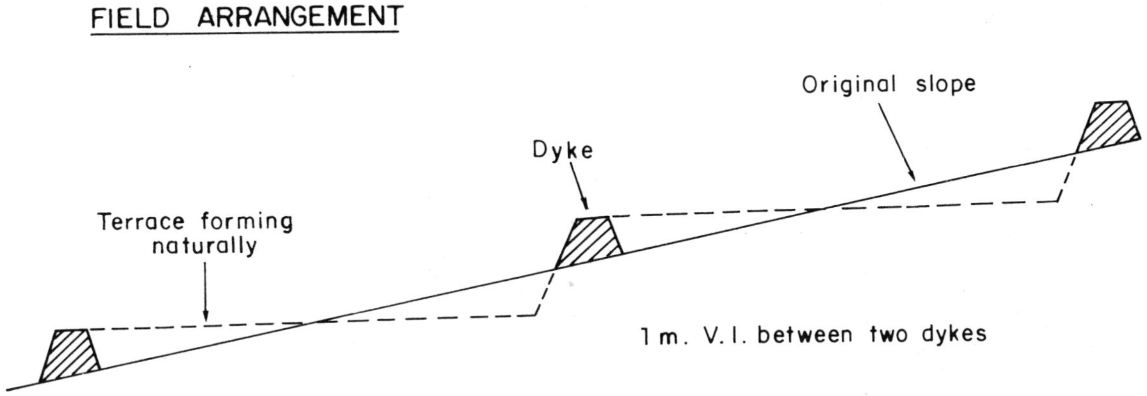

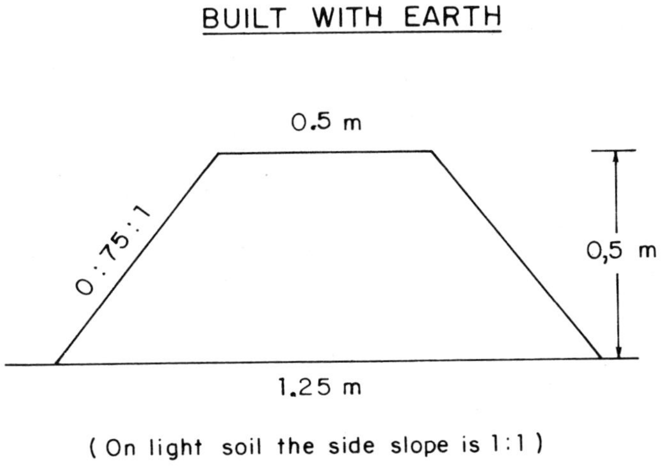

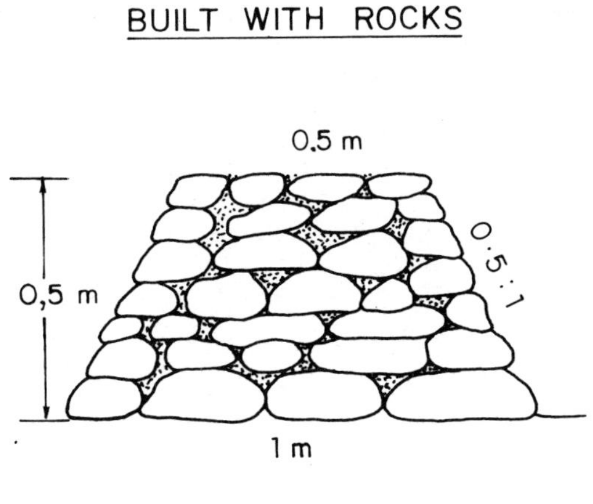

Fig. 14 Natural terrace design and cross-sections

years of cultivation. This land treatment for small farms replaces the broadbase terraces which have been used extensively on large farms in the temperate zones. Although they can be constructed on steeper slopes, the best sites for this land treatment are slopes of not more than 7° or 12 percent, and with soils of fair to good infiltration rates. A hectare of earth dykes may need 80 mandays to complete.

4.2.4 Waterways

As mentioned before, runoff is inevitable on cultivated slopes in the humid tropics. Wherever it has been concentrated, a safe disposal system is needed. Neglecting this basic fact will cause severe gully erosion. Depending upon the slope and the amount of runoff, natural depressions may not always be relied upon to carry additional runoff.

Man-made waterways usually form an integral part of terracing where runoff is inevitable. On steep slopes runoff may exceed 2 m/s in the waterways and under these conditions even a dense grass cover may not be enough to prevent scouring. In addition to this the flows must be concentrated as small farmers are usually not willing to give up strips of more than about 2 m wide. Under these conditions structures are needed to stabilize the waterways. Unfortunately, structures can be expensive but to avoid needless expenditure the following principles can be followed:

- if possible, divert water rather than concentrate too much runoff in one waterway;
- use adjacent grassland or forested areas as a water spreading area or a sump if these are available;
- select gentle slopes for waterway sites;
- use local material for building structures;

Figure 15 shows seven major types of waterways and their structural needs. Their uses and limitations are shown in Table 9. Small farmers considering future mechanization should use Type 5 waterways bearing in mind that terraces are permanent structures. Type 3 waterways need to be ballasted only in the middle, whilst the concrete structures used in Type 4a waterways can sometimes be replaced by masonry work if local materials are available.

Table 10 also lists several important equations used in estimating rainfall intensity, peak runoff, flow velocity and cross-sections of waterways for ready reference. Details and examples of the use of these equations for waterway design can be found in the FAO Watershed Field Manual: Slope Treatment Measures and Practices (FAO 1988).

4.2.5 Gully Control Structures

Small gullies can sometimes be erased by filling or reshaping and replanting vegetation provided that any runoff from above is safely diverted to another place. If these methods are not feasible, engineering structures become necessary to prevent further erosion or cutting of the channels.

Some of the waterway structures explained previously can be used in small gullies for erosion control purpose. For large gullies, however, different structures are usually required. Structures are expensive. Their cost should be paid for by government on community lands and shared with farmers on private farms.

The main types of structures needed for gully control are the following:

- for gully head protection: diversions, drop structures and cantilever chutes;
- for channel stabilization: check dams and submerged dykes;

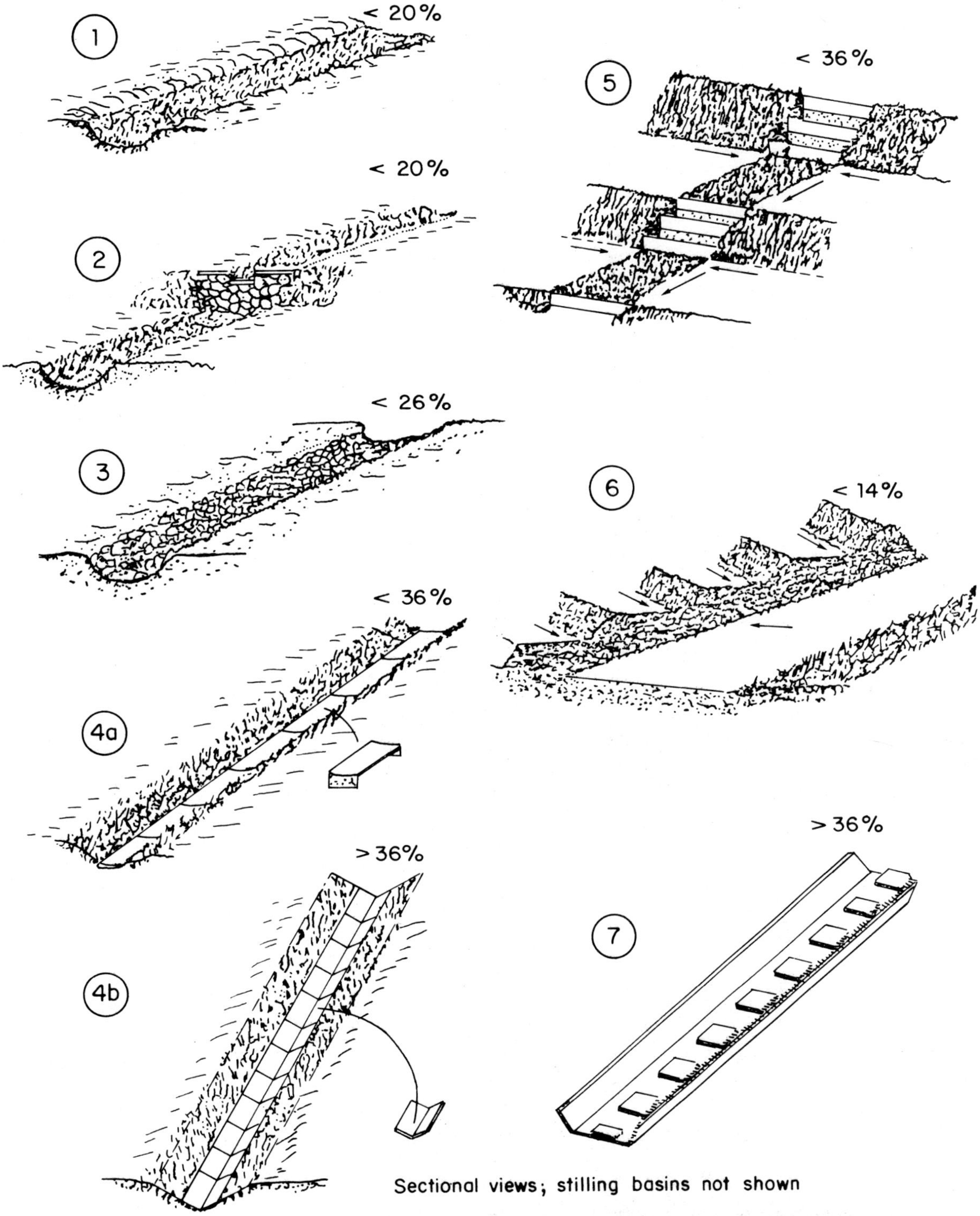

Fig. 15 Major types of waterway
(details explained in Table 9)

Table 9

MAJOR TYPES OF WATERWAYS: THEIR USES AND LIMITS*

Type	Shape	Channel protection	Velocity limit	Slope limit	Uses
Grassed waterway	Parabolic	By grass	1.8 m/s^{-1}	<11° (20%)	For new waterway or depression
Grassed waterway with drops	Parabolic	By grass and concrete or masonry	1.8 m/s^{-1}	Between two structures: 3%, overall slope <11° (20%)	For discontinuous type of channel
Ballasted waterway	Parabolic	By stones or stones in wire mesh	3 m/s^{-1}	<15° (26%)	Where stones are available
Prefabricated:					
(a) Parabolic waterway	Parabolic	By concrete structures and grass	-	<20° (36%)	A stilling basin is usually needed where rainfall is frequent and flows are constant
(b) V-notch chute	90° V-notch	By concrete structures and grass	-	>20° (36%)	Same as above and on very steep slopes
Stepped waterway	Parabolic and rectangular	By grass and concrete or masonry drops	On grass part: 1.8 m/s^{-1}	Overall slope <20° (36%)	For 4-wheel tractors and in the middle of bench terraces
Waterway and road ditch	Parabolic	By grass and stone ballasting	3 m/s^{-1}	<8° (14%)	For 4-wheel tractors; mechanization
Footpath and chute complex	Trapezoidal or rectangular	By concrete or or masonry structures	-	>20° (36%)	For paths on small farms and on very steep slopes

* These limits are approximations for general reference. In practice, the volume and velocity of runoff and site conditions should all be taken into consideration.

Table 10 BASIC EQUATIONS FOR WATERWAY AND STRUCTURE DESIGN

Estimation of rainfall intensities using maximum daily rainfall*:	$I = \frac{R}{24} \left(\frac{24}{t} \right)^{0.6}$	I: rainfall intensity, in mm/hour R: maximum daily rainfall of 10 year frequency, in mm t: time of concentration, in hour
Estimation of peak runoff using "Rational Formula"	$Q = \frac{1}{360} C\ I\ A$	Q: peak runoff, in m^3/s C: runoff coefficient, the percent of rainfall appears as runoff I: maximum rainfall intensity for a given frequency in a duration equal to time of concentration, in mm/h A: area of watershed, in ha
Estimation of flow velocities using "Manning Formula":	$V = \frac{1}{n} R^{2/3} S^{1/2}$	V: average velocity of flow, in m/s n: roughness coefficient of the channel R: hydraulic radius equal to the cross-sectional area divided by the wetted perimeter S: hydraulic gradient (channel slope), in percent
Estimation of waterway cross-sections:	$A = \frac{Q}{V}$	A: cross-sectional area, in m^2 Q: discharge, in m^3/s V: velocity, in m/s
Estimation of cross-sectional area of a parabolic waterway:	$C = \frac{2}{3} T\ D$	C: cross-sectional area, in m^2 T: top width of waterway, in m D: depth of waterway, in m

* This equation has been used in tropical countries where only daily rainfall records are available.

i. Diversions

Where practical, runoff should be diverted from above the head of a gully to a safe disposal point. A diversion can be a hillside ditch (see 4.2.3), a trapezoidal trench or a contour dyke. They should be large enough to carry a 10 year frequency storm runoff and built on firm ground and at safe distance from the gully head.

ii. Drop structures

Drop structures are often needed to prevent the vertical gully head from cutting upslope in sites where the diversion of total runoff is impractical. A drop structure, as used in a waterway, is composed essentially of: (a) a head wall with extensions and a weir, (b) two brace walls, (c) cut-off walls underneath, and (d) an apron or a stilling basin with a sill. The head wall should be built against the gully head with a weir at its top. The flow capacity of the weir is determined by the formula: Q = AV (see Table 10). The apron or stilling basin should be well built and long enough to eliminate turbulence, usually 1.5 to 2 times the height of the head wall (up to the height of weir). A diagram of a drop structure is shown in Figure 16.

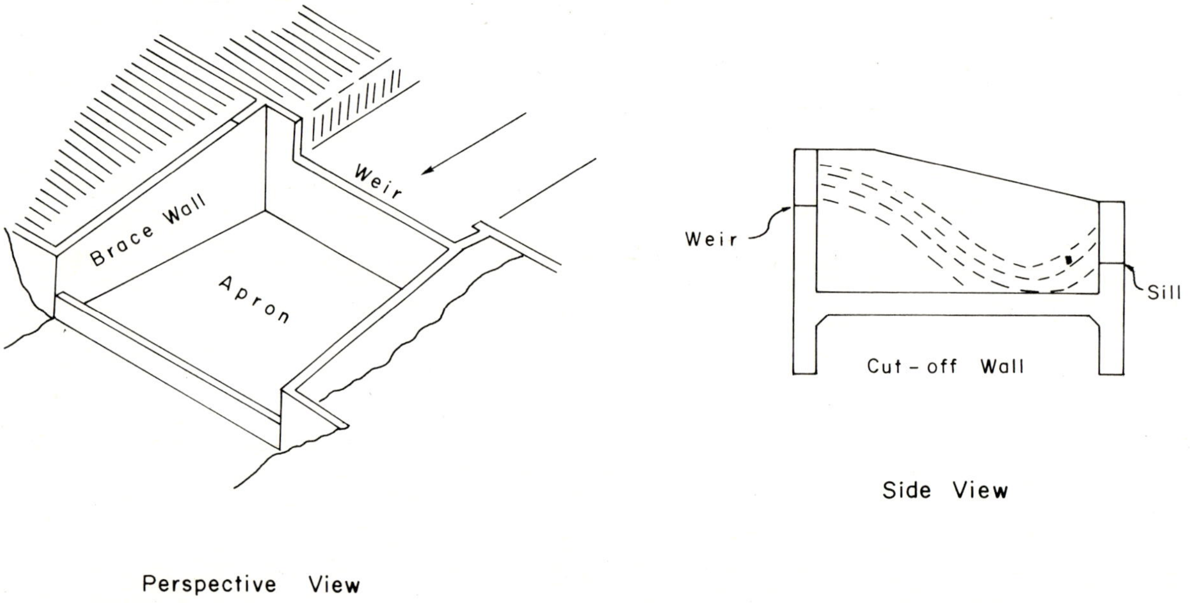

Fig. 16 Diagram of a drop structure

iii. Cantilever chute

A cantilever chute is used where the gully head is very deep, and the runoff cannot be diverted to another place. The structure usually consists of: (a) a trough to receive the runoff from the area above, (b) a supported concrete channel which carries the runoff to a stilling basin to dissipate the energy of the falling water. This structure is used to drain large quantities of runoff and to protect highly productive land from being eaten away by a gully.

iv. Check dams

Check dams are used in gullies and upstream channels to:

- check or slow down flow velocities;
- arrest sediments eroded from the land or channel above;
- confine flow alignment so as to prevent bank cutting;
- reduce channel gradients after silting up.

Check dams can be of a temporary or permanent nature depending upon actual needs and the material to be used. Temporary ones are used to slow down the flow and arrest the silt until enough vegetation is established to protect the channel. They are used mostly in small gullies in semi-arid regions where runoff is occasional and in relatively small quantities. Brush dams, log dams and loose rock dams etc. are normally used for this purpose depending on which local material is readily available. However, in the humid tropics where frequent and large quantities of runoff prevail, permanent structures are needed. Brick, masonry and concrete dams belong in this category. If no gravel is coming down from the watershed above, rust-resistant wire mesh gabions can be used effectively for making check dams.

The components of a check dam are similar to a drop structure, except that the brace walls and sill are omitted in most dams. Its wing walls should be keyed deeply into the gully banks. The wing walls should also be sloped toward the weir to confine the flow to the middle in case the quantity of actual runoff is greater than calculated. For gullies on farm land, check dams lower than 2 m are usually recommended. Dams over 2 m should be considered as major engineering works and civil engineers should be consulted.

In designing a series of check dams the following principles should be observed:

- A particularly stable site should be selected at the mouth of a gully and a strong dam built to avoid a domino effect to the upper dams in case of failure. The channel slope below the dam should be nearly flat or a gentle slope so that no serious undercutting is likely to occur.

- The height of the dams should be planned in such a way that the future channel gradient will be as near to level as is feasible, keeping in mind that the channel will eventually be protected by grass. The gradient can be determined by the "Head to toe relationship" illustrated in Figure 17.

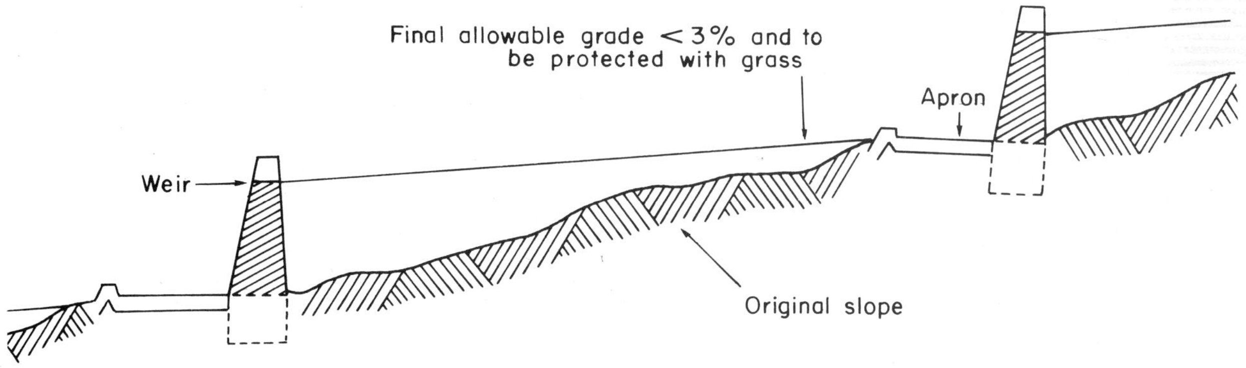

Fig. 17 Example of check dams. Gradient between head of lower weir and toe of next upper weir not to exceed 3 percent

- The horizontal distance between dams can be determined using the following equation:

$$L = \frac{100}{M - N} H$$

where L: horizontal distance between dams
M: original gully gradient
N: proposed gully gradient after sediment deposition
H: dam height up to the weir

- Check dam sites should normally be selected at narrow places with large storage spaces behind the proposed dams. Therefore, some modification of the horizontal distance is allowable, or adjustment of heights, to get the desirable gradient.

- If no drop structure is built at the gully head, a taller and stronger check dam should be constructed not far from the head. The height of this dam should be such that, when it is filled with silt, the gully head will be covered.

- Check dams should be placed with all the weirs, or the line of discharge, coinciding with the centre line of the channel in order to minimize the danger of bank cutting.

- The weir of a check dam should be designed to handle a 10 year storm frequency for ordinary cases and 25 year frequency or more for special cases.

v. <u>Submerged dykes</u>

Submerged dykes are low walls built across a channel to prevent down-cutting. Alternatively they can be used immediately downstream from check dams to prevent scouring. In either case, the dykes should be keyed firmly into the channel and the two banks. The tops of the dykes should be designed in uniform grade downstream to control the channel gradient.

4.3 AGRONOMIC CONSERVATION MEASURES

Agronomic conservation measures are, in essence, proven techniques of soil, crop and livestock management, closely related to normal farming practice, but specially designed or selected to be beneficial in terms of conservation. They are usually inexpensive and can be adopted relatively easily by small farmers. However, before such measures are introduced to a new locality, their compatibility with existing farming methods and cropping systems should be carefully considered, together with their likely effectiveness in erosion control under the conditions prevailing.

On steep slopes in the tropics, agronomic conservation measures should be used in conjunction with conservation structures for, used alone, they may not be totally effective. They have proved most effective on gentle slopes below 12 or 15 percent (see 4.3.4 etc.) but when used in conjunction with discontinuous types of terrace, for example, they allow safe cultivation of steeper slopes than would be possible in their absence. Plates 5 to 8 show agronomic measures in use with simple structures.

4.3.1 Multiple Cropping

Multiple cropping, in general, can be defined as growing more than one crop on the same piece of land in one year. The system can take various forms, including intercropping, relay cropping, and mixed cropping (Harwood 1979). Apart from increasing crop production, all forms can be beneficial to erosion control because they add an earlier or later protective cover to the ground (when compared to monocropping). From an erosion control viewpoint, special attention should be paid to those crops which can provide dense and low ground cover or thrive before heavy rains. Yet when interplanted, they should not seriously affect the yield of other crops.

Multiple cropping is being practised by small farmers in many parts of the world. There are many examples in Indonesia, for instance, where small farmers are intercropping upland rice with maize, maize with cassava, maize with groundnuts and sorghum with millet. In the rice-maize intercrop, the maize grows more rapidly than the rice and is harvested before the heading of the rice plants. The growth of the rice is only slightly retarded by the maize, and later proceeds to maturity giving a relatively high yield. In research findings, these intercrop combinations have yielded up to 60 percent more than monocultures (Harwood 1979). At other places in Southeast Asia, maize is planted in rows 2 to 3 metres apart and intercropped with soybean, groundnuts or mungbean. Not only do the legumes provide nitrogen to the soils but also the maize plants suffer less from pests and disease. In Latin American countries, maize intercropped with beans is very popular. Between rows of maize, beans can thrive, covering the ground with a dense mat of vegetation (Plate 5).

Plate 5 Maize intercropped with beans on a gentle slope supported by contour dykes. Natural terraces are gradually forming after a year of cultivation (El Salvador)

The Tropical Agricultural Research and Training Center (CATIE) in Costa Rica has developed many multiple cultivation systems for small farmers. Also in Africa, small farmers traditionally practise intercropping, and monoculture has never been common (Ofori 1974). There, land freshly cleared for maize may be planted with other crops such as plantain, cocoyam and eventually cassava. Although the system mainly provides insurance against crop failure, it also reduces soil erosion. In the last decade, the International Institute of Tropical Agriculture (IITA) in Nigeria has conducted research on these effects.

However, research findings on the use of multiple cropping as a protection against soil loss and runoff are still rare. Data applicable to steeply sloping land are especially scarce. Soil conservationists in developing countries need to collect local information and to develop systems which will benefit both small farmers and their land.

4.3.2 Contour Cultivation and Close Planting

For erosion control, contour cultivation is better than the up-and-down slope tillage method still used in some countries. However, merely planting crops along the

Plate 6 Contour beds built by a small farmers on 15 degree slope. Note hillside ditches are built to support the contour beds (Jamaica)

Plate 7 Contour and close planting of pineapples with partial mulching on a 25 degree slope. These agronomic conservation measures are supported with hillside ditches (Jamaica)

contour will not significantly reduce soil loss and runoff. Close planting on the contour will yield better results whilst planting crops on contour beds controls erosion and runoff even better.

Unfortunately contour beds are difficult to build and maintain precisely on the contour (or graded contour), particularly on uneven or dissected terrain. After heavy rains, water will overflow at any low spots and gullies can quickly develop. Experience shows that contour beds are more suitable on slopes below 7° or 12 percent and on permeable soils. When they are built on steep slopes or on heavy soils, they should be protected by hillside ditches and waterways. Plate 6 shows contour beds built by a small farmer in Jamaica on a 15° slope supported by hillside ditches. Contour beds can also be costly to install on steep slopes. The labour used to build them per hectare every year for 3 years is almost equal to the cost of building a hectare of narrow bench terraces which are more permanent. However, contour beds are needed to grow many kinds of root crops successfully, even on bench terraces.

Close planting has the benefit of leading quickly to a dense ground cover and of retarding soil loss. For many crops, close planting will also increase production. For instance, it has been found in Taiwan that raising the pineapple population per hectare from 25 000 to 45 000 combined with contour and close planting and partial mulching not only reduces erosion to almost the same level as on bench terraces, but also produces the best yields (JCRR 1977). In this instance, however, erosion increased significantly at the end of the rotation, when pineapples were discontinued. Recently, a research report from Trinidad (Gumbs **et al.** 1985) stated that increasing the maize population from the usually recommended 41 500 plants per ha to 62 500 plants per ha, has resulted in less soil loss (20.5 t vs. 32.0 t) and greater fresh weight yield of ears (5.1 t vs. 4.2 t). Close planting could be even more effective in erosion control if supported with simple structures.

4.3.3 Strip Cropping

Strip cropping is a conservation practice in which crops are grown in a systematic arrangement of strips or bands that serve as barriers to water and wind erosion. There are three major types:

- Contour Strip Cropping: Two or more crops are planted along contours in alternate strips.
- Field Strip Cropping: The alternate strips are of a uniform width across the field and do not necessarily curve to conform to the contour. This is used on very irregular topograpy.
- Wind Strip Cropping: Tall, wind-resistant crops and normal crops are planted alternately in narrower strips perpendicular to the direction of prevailing winds.

All these types have a strip of an erosion-resistant crop, either for wind or water erosion, planted alternately with another strip of clean cultivated row crop. Crop rotation can be practised by strips. The width of the strips is determined by site conditions.

Strip cropping is effective for erosion control under mild rainfall and on gentle slopes (up to 7° or 12 percent, or a little more). It is more practical on large farms and uniform topography than on small and dissected sites. For steep small farms in heavy rainfall regions, it should be supported with structural measures such as hillside ditches or contour dykes, etc. (see Plate 8). For farmers to accept these practices, the erosion resistant crop, grass or legume, should yield an additional benefit to them, e.g. forage, seed or food.

Plate 8 Strip cropping of maize, upland rice and groundnuts is supported with contour dykes (mid-picture) whilst contour and close planting of beans with hillside ditches can be seen in the foreground (Northern Thailand)

4.3.4 Mulching

Mulching is quite effective for erosion control because it protects at ground level, forming a cover against splash erosion and capping. It is also a practice particularly beneficial to tropical soils (ter Kuile 1984). Mulching can: (i) provide a continual supply of organic matter to replace that lost by microbiological processes in the soil; (ii) protect the soil from extreme temperatures and allow the soil fauna to continue their useful work; and (iii) maintain the exchange capacity at a level where nutrient leaching losses are minimized and hydrogen saturation (acidity) is kept within bounds.

Research conducted in Nigeria (Lal 1978) and Indonesia (Abujamin 1985) has found that mulching plots (up to 15 percent slope) considerably reduced soil loss and runoff. The Indonesia study found that mulching and minimum tillage on a 14 percent slope could reduce soil loss and runoff more than 90 percent in comparison with the control plots.

Materials used for mulching can be crop residues, straw, green leaves and stems, and even gravel etc. as available to small farmers. Mulching can be applied over the whole field; between crop rows or along the contour; or around single plants in the case of tree crops. Mulching used as an agronomic measure can considerably reduce weeding costs. However, if the material has to be collected and hauled from another place, the transport cost can be prohibitive; up to 10 tons of material may be initially needed for one hectare, and it requires annual replenishment. Farmers who practise mulching should be encouraged to grow their own material or practise live mulching and minimum tillage techniques (see 4.3.6).

There are two problems with mulching. One is the cost and the other is its doubtful effectiveness on steep slopes when used alone. Simple structures may still be needed for support (see Plate 7).

4.3.5 Cover Cropping

Cover crops are close-growing crops planted mainly to protect the soil between tree crops or semi-permanent crops, or, between the seasons for regular crops. The types of cover crops can be grasses or legumes, annual or perennial, depending on the actual needs.

Generally speaking, legumes can provide ground cover and add nitrogen to the soil at the same time. However, some aggressive types of legumes such as tropical kudzu **(Pueraria phaseoloides)**, centrosema **(Centrosema pubescens)** and desmodium **(Desmodium** spp.) etc. used in banana or citrus orchards will create extra labour to keep them away from the main crops. Small farmers may not always welcome them, but they can be used between periods of main crops and during fallow as a cover crop or green manure. Erect or prostrate type of perennial legumes such as stylo (**Stylosanthes** spp.), Lototonis **(Lototonis bainesii)**, creeping indigo **(Indigofera endecaphylla)** etc. may be more suitable for use in orchards. Small farmers sometimes use leguminous cash crops for temporary cover on the ground until fruit trees or bananas etc. are established. Groundnuts, cowpeas **(Vigna sinensis)**, and many other kinds of legumes can be used, but because of their short lifespan and the frequent disturbance of the soil, they should be planted with caution.

Grasses can also be used as a ground cover for orchards or planted under semi-permanent crops. Local grasses can be easily grown for this purpose. The best ones should be shade tolerant, easily propagated and require least management. Bahia grass **(Paspalum notatum)**, for instance, has been successfully and extensively used as a cover crop in orchards. Table 11 lists some useful plants for soil conservation in the humid tropics, including cover crops.

Table 11 SOME USEFUL PLANTS FOR SOIL CONSERVATION IN THE HUMID TROPICS

Botanical name	Common name	Major use*	Remarks
Legumes			
Calopogonium mucunoides	Calopo	C mainly	Perennial
Canavalia ensiformis	Jack bean	C & G	Annual
Cajanus cajan	Pigeon pea	C & G	Annual
Centrosema pubescens	Centrosema	C mainly	Perennial
Crotalaria juncea	Sun hemp	G mainly	Annual
Desmodium intortum	Greenleaf Desmodium	C & G	Perennial
Indigofera endecaphylla	Creeping indigo	Cover crop	Perennial
Lablab purpureus	Lablab bean	C & G	Annual
Leucaena leucocephala	Leucaena	E & M	Perennial
Lotononis bainesii	Lotononis	E & C	Perennial
Lupinus luteus	Yellow Lup.	C & G	Annual
Mucuna capitata	Velvet bean	C & G	Annual
Stylosanthes guianensis	Stylo	C & G	Perennial
Vigna sinensis	Cowpea	C & G	Annual
Grasses (perennials)			
Axonopus affinis or **A. compressus**	Carpet grass	E mainly	For risers & waterways
Brachiaria mutica	Para grass	E mainly	Waterways
Cynodon dactylon	Bermuda grass	E mainly	For road
Digitaria decumbens	Pangola grass	E mainly	Good forage
Eragrostis curvula	Weeping lovegrass	E mainly	Road etc.
Melinis minutiflora	Molasses grass	E & C	Good for both
Panicum maximum	Guinea grass	E mainly	For risers
Paspalum notatum	Bahia grass	E & C	Good for both
Pennisetum purpureum	Napier grass	M mainly	Good fodder
Pennisetum clandestinum	Kikuyu grass	E mainly	High elevation

* C: Cover crop; G: Green manure; E: Erosion control; M: Mulching

With dense cover and roots, cover crops can be very effective in erosion control. Research in various orchards on slopes ranging from 14 to 46 percent in Taiwan showed that the effectiveness of cover crops (using their foliage for mulching) in erosion control was on average 95 percent, with Bahia grass the most effective (Wu 1986). A runoff plot study in Jamaica showed that Bahia grass cover in bananas reduced erosion about 83 percent on a 17° or 30 percent slope.

However, taller crops (0.5 m) with big leaves were found to increase erosion. This may be due to the increase in falling distance and drop size of the intercepted rain, therefore more research is still needed on the effectiveness of cover crops (Morgan 1984). In addition, cover crops may compete for moisture and nutrients and influence yields of the main crops. Special management such as timely cutting of foliage, circle weeding and probably more fertilizer inputs are needed. On the other hand, simple structures such as orchard terraces, individual basins and/or hillside ditches may still be needed before dense cover is established to reduce erosion. In fact some of these structural measures are used not only for erosion control but also to facilitate crop management.

4.3.6 Minimum Tillage

Minimum tillage is defined as the minimum soil manipulation necessary for crop production. A broader term, conservation tillage, is sometimes used to emphasize the kind of tillage system that reduces loss of soil and water (Mannering and Fenster 1983).

Minimum tillage in practice is to till only the crop zones for seed planting, germination and crop establishment. Because it disturbs soils as little as possible, erosion is minimized. The practices may involve mulching with crop residues, use of special equipment and herbicides etc., depending on various conditions. These kindsof practices have been applied extensively in countries like the USA and Canada. They have not yet been developed enough to be recommended for small hill farmers in the humid tropics. Foreseen difficulties may be as follows:

- small farmers may be prevented from applying the necessary techniques, e.g. using chemicals or special tools, by their limited resources;
- crop production per unit area may be lowered and small farmers have no other areas on which to make up the losses;
- root crops are the small farmers' staple food in many parts of world; digging, working or piling the soil (in the case of yams) makes minimum tillage difficult;
- research results for small farmers are still sketchy.

IITA in Nigeria is one of the few institutes that has conducted some research in this respect. One approach tested was mulch tillage (a kind of minimum tillage) using either live mulch such as low growing legumes or crop residues **in situ**. The mulch tillage was based on the zonal tillage concept, whereby the seed bed was divided into a seedling zone and a soil management zone. The seedling zone served to provide the ideal conditions for seed germination and establishment whereas the interrow soil management zone served to conserve soil moisture, prevent runoff and erosion and optimize the soil temperature regime. It was reported that this method could minimize soil erosion, maintain soil structure and produce yields equal to ploughed plots in the long run though it might not yield the maximum (Lal 1978).

A soil loss study carried out in northern Thailand showed that upland rice planted on a 30 percent slope with minimum tillage as control (dibbling seeds only and no tillage) had a soil loss of 24 tons per hectare per year, whereas the usual clean cultivation of upland rice on the hills of Northern Thailand could entail losses of 50 to 100 tons/ha/year (Wichaidit **et al.** 1977; Marston **et al.** 1985). When the dibbling methods were supported with simple conservation structures, the erosion rates were found to be around 12 t/ha/yr, another 50 percent reduction (Sheng **et al.** 1981).

4.3.7 Vegetative Barriers

Vegetative barriers are bushes or hardy grasses planted across a slope in order to retard the soil from being washed downslope. They are of two general types. One approach is to plant the barriers at an appropriate interval solely for erosion control purposes (using for example the same distance calculated for hillside ditches). Below one or two barriers, contour drains are sometimes installed to control excess runoff. The other approach is to plant the barriers much closer, usually spaced according to the future bench width and similar to the layout of natural terraces explained previously (see 4.2.3). After several years of cultivation, rough bench terraces can be expected to form.

The barriers can be one row or several rows and of different widths according to the needs and the farmer's acceptance. Easily propagated, local species can be used for this purpose. Terrace-forming types, however, should be tough, hardy and thick growing, and likely to sprout again if buried.

Both approaches can be effective where the slopes are gentle and the rain not very intense. Careful management of barriers is essential to obtain the expected results. For instance, grasses should be tended and cut back to avoid over-shadowing the main crop or growing aggresively into the crop zones. The foliage obtained can be used for mulching.

To use vegetative barriers on steep slopes in regions with heavy rainfall, without the support of conservation structures to take care of excess runoff, will often result in failure. Neither in Jamaica nor in El Salvador has the erosion control function of vegetative barriers on steep slopes been successful, although barriers have been tried over decades. Research conducted on a 32 percent slope with upland crops between grass barriers in Taiwan showed a total soil loss of 273 tons per ha in two years (Hsu **et al.** 1977).

Many agro-forestry practices also involve interplanted trees or grasses either in rows across the slope or in other forms to minimize erosion. A popular practice is to plant Leucaena as a quick-growing hedge along the contours to hold back eroding soil from the cultivated strips above. Simple structures are still used to safely drain away the runoff (see 4.5.3).

4.3.8 Partial Weeding

Partial weeding not only benefits erosion control, but also saves labour. Circle weeding around fruit trees or bananas is usually sufficient, especially when the trees or plants are planted in individual basins. For plantations of tea and coffee, etc. strip weeding can be applied, leaving the interrow spaces as a grass buffer. Occasionally the grass should be cut low, but there is no need to dig it out.

The problem with partial weeding is that many farmers have been practising clean weeding for so long that as a practice it cannot be accepted easily. Some weeds may also harbour insects and pests. Farmers therefore need time to find the real benefits and become convinced.

4.4 SOIL MANAGEMENT

Soil management is very necessary for many small farms in the humid tropics where the soils have been degraded through decades of abuse. During land treatment by structural means, subsoils may also be exposed.* Unless soil improvement measures are followed the productivity of the treated lands will be low. Good soil management practices will not only improve or maintain soil structures, aggregate stability, infiltration capacity and benefit erosion control, but will also increase soil productivity. The following sections are brief descriptions of some major measures which small farmers can adopt.

---

* This can be avoided only by stockpiling and returning the topsoils.

4.4.1 Deep Ploughing

Deep ploughing is necessary on newly-built bench terraces where deeply cut areas (see Figure 7) are not ready for planting crops. It also applies to heavy soils or compacted lands. Because it disturbs the soil and increases the erosion hazard, this practice should only be applied where the land is protected by conservation structures.

Its main objectives are: (i) to provide spaces for air and water in the soils to benefit crop growth; (ii) to develop better soil structures, especially with added organic matter; (iii) to promote earthworm and microbiological activities; and (iv) to break soil crusts, hardpans, etc. and to increase the infiltration capacity.

The depths needed for such ploughing depend on the soil, crop requirement and the tools to be used. For deep rooted crops like yams (**Dioscorea** spp.) on a newly-built terrace of clayey soils, for instance, a depth of 45 cm (18 in.) is needed. Other crops on silty soils may need much less.

4.4.2 Use of Compost and Farm Manure

Compost and farm manure etc. add organic matter to the soil to improve both soil fertility and soil structure. In the humid tropics where oxidation and leaching processes are much more active than in other climatic zones, the organic content of exposed soils quickly becomes low and often needs to be replenished.

Small farmers can make compost in pits or piles in the open air. The site should be near enough to the area of application to reduce future transportation chores. Usually, the compost is built by putting down alternate layers of plant material and farm manure. The whole mass should be kept moist in order to reduce the loss of ammonia and elemental nitrogen, and also to encourage vigorous bacterial action. Sometimes, phosphate is also added to balance the mixture and make it more effective. Regular turning is needed to maintain aeration and promote microbial activities. Attention should be taken not to let too much rain water soak in, and any solution from the compost should be collected and eventually used for fertilizing.

In many countries, farmers apply farm manures and stable material directly onto the field. This may be due to an abundance of manure, lack of suitable plant material, or a shortage of labour. However, farmers should be encouraged to use only well-rotted manure for this purpose. Also they should be encouraged to collect any kitchen remains such as vegetable peelings, fish bones, feathers and other garbage for soil improvement purposes.

Since compost and manure need considerable labour and time to make or collect, and the required quantity is large, farmers should be encouraged to use them more selectively, e.g. on terraces for vegetables and cash crops, or on newly reclaimed fields.

4.4.3 Green Manuring

It is a common practice to turn under green plant material to improve the soil. Green manure usually provides an inexpensive and convenient source of organic matter for replenishing plant nutrients, improving the soil structure and minimizing runoff and erosion.

Many leguminous crops are used for this purpose. Their roots add considerable nitrogen to the soil, and they take up nutrients from within the soil which would not otherwise have been available for future crop use. If the land has not been previously used for growing leguminous crops, seed inoculation is advisable.

Many legumes mentioned in section 4.3.5 and listed in Table 11 can be used as a green manure. Cover crops which were planted between the two main crops can be ploughed in at the season's end as green manure for the following main crop.

Green manuring is most needed on newly constructed terraces or reclaimed lands when other sources of organic matter are not available. Sometimes it is used to supplement compost and farm manure which are most labour-consuming. Generally speaking, the best time for turning the plants in is when the crop is in flower.

For some small farmers, to plant a crop only to be ploughed in is a luxury they cannot afford. Also, ploughing in needs equipment and time. Therefore they should be encouraged to leave as much foliage and residue of the main crops in the field as possible and not to burn them.

4.4.4 Use of Fertilizers

Proper application of fertilizer will not only improve soil fertility but also promote faster crop cover to protect the ground from splash or wind erosion. The soils of many upland farms have been eroded and severely degraded. On these soils, and on newly-reclaimed land or new terraces, proper fertilizer application is usually required in order to restore their productivity.

The kinds and amounts of fertilizer to be used depend upon the different soils and crops. Nitrogen is usually the main element needed, while phosphorus and potassium are frequently required. Lime may be needed on acid soils. Soil tests, fertilizer trials and keen observations will help to determine the appropriate kinds and amounts of fertilizer to be used.

Since fertilizer prices are relatively high, their application on small farms should be as cost effective as possible. The following guidelines for use are suggested:

- use fertilizers on bench terraces or nearly flat lands to minimize the chances of their being washed away;

- carry out fertilizer dressing properly; cover with soil, do not put on the top of high mounds or expose to wind or rain;

- use fertilizers for high-price cash crops, for newly-reclaimed areas or land being protected by conservation measures;

- use fertilizers in conjunction with compost, farm manure, and green manure to obtain better results and reduce the fertilizer inputs needed.

In many countries, subsidies and credit may be available for small farmers to use fertilizers. Even so, economic use of fertilizers is imperative. Improper use, including overdose, is not only a waste but can be detrimental to the land and the environment.

4.4.5 Crop Rotation

Crop rotation is the growing of different crops in recurring succession on the same piece of land in a planned cycle. The crops used in rotations normally include row crops, small grains, legumes and grasses or their mixtures. The inclusion of legume/grass mixtures in the rotation will not only enrich nutrient reserves, but also improve soil structure. The best system will depend on the farmer's preference, economic advantages, and physical environment.

Crop rotations benefit soil improvement, conservation and production. Continuous monocropping generally depletes soil nutrients and increases pest and disease hazards. Rotations also give small farmers a variety of crops for diverse uses as well as a flexibility in marketing.

An early report from Nigeria (Vine 1953) stated that a long-term rotation of maize and mucuna (a legume used as a cover crop) was able to maintain maize yield for 20 years without fertilizer. For those small farmers who cannot afford the use of fertilizer, crop rotation may be quite useful to maintain a moderate level of productivity.

The types of crops that can be grown in a rotation are largely influenced by rainfall, labour supply and marketing. With a reasonable supply of moisture in the soil as well as adequate labour, small farmers are more likely to accept pulses and vegetables in the system rather than grasses or cover crops. In the upland of northern Thailand, for example, upland rice followed by groundnuts, taro and then soybean; or, maize followed by mungbean, taro and then kidney bean are becoming popular rotations. In southeast Jamaica, onion, groundnut and green maize; or red pea, cabbage, and cucumber; or sweet potato, cowpea and carrot, have been suggested to the small farmers by an FAO project. Each of these rotations includes legumes and vegetables.

4.4.6 Soil Moisture Conservation

Rainfall distribution can be quite irregular in the humid tropics. Whether they have one or two peak rainfall seasons, many countries in the humid tropics have a dry season with several months of low rainfall. During this dry period, many lands are unused and left bare, or the crops are scanty. If moisture could be made available to grow a good crop during this period, this would greatly benefit the economics of small farming.

Soil moisture conservation practices include mulching, use of organic matter, levelling or bench terracing, etc., or their combinations. A study in Taiwan found that the soil moisture content on bench terraces is much higher than that on slopes (Chiang 1965). Mulching reduces the evaporation of soil moisture considerably. Soils with more organic matter usually hold soil moisture effectively and are more resistant to minor droughts.

The technical aspects of mulching, composting and terracing have been discussed previously. Some additional practices for moisture conservation are briefly introduced below:

- observe a rule of 'no burning' during the entire dry season. If fire is necessary to clear the fields, the litter or weeds should be concentrated and the fire be confined to a small area;

- no tillage of soils should be practised as the dry season approaches;

- establish shelterbelts, permanently or temporarily, perpendicular to the prevailing wind direction where strong, drying winds prevail;

- harvest and store runoff during the rainy season to be used as supplementary irrigation during the dry period. For example, collect runoff from roads, waterways, or gullies and store it in tanks or ponds for use by crops.

4.5 INTEGRATION WITH FARMING SYSTEMS

For soil conservation to be successful, it should be well-integrated with the prevailing farming systems and become an indispensable part of the overall package of measures being advocated. Many conservation projects have experienced problems because soil conservation has been presented as an extra task. If soil conservation can be rooted firmly in the farming system, the work will be carried out more willingly and smoothly.

In the following sections, three major land use systems are briefly discussed with emphasis on soil conservation. All the systems incorporate erosion control measures which will benefit small farmers as well as the upland environment.

4.5.1 Hillside Farming and Conservation Farming Systems

Any farming system practised on hillsides in the humid tropics should take into account the following problems:

- problems of rapid erosion caused by heavy and intense rainfall, quick runoff and steep topography;

- problems of rapid soil degradation caused by high temperature, quick decomposition of organic matter, and leaching of nutrients;

- problems arising from traditional but inappropriate ways of land use and tillage which accelerate the rate of land degradation;

- other social, economic and infrastructure problems bearing on wise land use, which may be very localized.

Many of these problems have been discussed in previous chapters and this section will focus on major techniques and approaches to counter them.

Firstly, hill farming systems should be based upon a land capability plan and land only used according to its capability. A farm plan should be prepared (see 3.3.3) to serve as the base for proper land use. Land use adjustments should be made to correct any existing misuse.

Secondly, the system should include conservation measures for each piece of land according to its erosion control needs. These might include major conservation structures and agronomic measures depending on farmer's interest, land capability and crop needs. Table 12 shows an example of planning used in Northern Thailand.

Thirdly, the system should include major farming and crop production activities with emphasis on rotations, soil cover, and the improvement of soil fertility and soil structure. Any farming practice which is destructive of land and environment should be discouraged. The farming system should also incorporate better farm management practices and infrastructure e.g. farm roads, storage, marketing, shelterbelts, effective drainage and, if possible and appropriate, minor irrigation.

Last but not the least, the system should include maintenance practices for both conservation and infrastructure.

If hill farming systems can adopt the above mentioned principles, then they can be considered as conservation farming systems.

Conservation farming is a system of farming which protects the land from any form of deterioration yet increases or maintains farm production at the highest possible level to the benefit of the farmers. The system, if carefully designed and implemented, should be complementary to modern farm management and supportive to the farmers' production goal.

Examples of conservation farming are not lacking. Many paddy farms in Southeast Asia are well protected and managed; tea, coffee, rubber, vegetable gardens and orchards in many parts of the world can also serve as demonstrations and focal points. The problem is that the vast uplands in the humid tropics under cultivation by small farmers are mostly under land destructive practices. Small farms in the uplands, which are by far the most numerous type in the tropics, have long been neglected by their governments. To assist them to establish conservation farming systems is an urgent task and a challenge to many developing countries.

4.5.2 Agro-forestry

Recognition of agro-forestry as a science is relatively new, although in practice it has been used for centuries in some countries. Agro-forestry covers a wide range of activities including agro-silviculture, silvo-pasture, agro-silvopasture and multi-purpose forest production (King 1980). In each of them, woody perennials (trees, shrubs, palms, etc.) are deliverately grown on the same piece of land with agricultural crops, fodders or animals (Nair 1983). The main objectives of agro-forestry are two-fold. One

Table 12 MAJOR CONSERVATION NEEDS FOR DIFFERENT CAPABILITY OF LANDS AND DIFFERENT LAND USE FOR FARM PLANNING IN NORTHERN THAILAND

Capability	Land use	Soil depths >90 cm	50-90 cm	20-50 cm	<20 cm
C_1 <12%	Annual crops	AC,CD,HD*	AC,CD,HD	AC,CD,HD	AC,CD,HD
	Semi-permanent crops	AC,HD	AC,HD	AC,HD	AC,HD
	Tree crops	-	-	-	-
	Pasture	-	-	-	-
	Forest	-	-	-	-
	Reserve or recreation	-	-	-	-
C_2 12-35%	Annual crops	BT,IT,HD	BT,IT,HD	HD	-
	Semi-permanent crops	HD&AC	HD&AC	HD&AC	-
	Tree crops	OT,CT,HD	OT,CT,HD	OT,HD	-
	Pasture	-	-	HD,DD	HD
	Forest	-	-	N	N
	Reserve or recreation	S	S	S	S
P <35%	Pasture	HD,DD	HD,DD	HD,DD	HD
	Forest	N	N	N	N
	Reserve or recreation	S	S	S	S
FT 35-50%	Tree crops	OT	OT	AF	-
	Forest	N	N	N	N
	Reserve or recreation	S	S	S	S
F 50%	Forest	N	N	N	N
	Reserve or recreation	S	S	S	S
RF	Reserve or recreation	S	S	S	S

* Major conservation needs:

AC: Agronomic conservation measures
CD: Contour dykes or natural terraces
HD: Hillside terraces
BT: Bench terraces
IT: Intermittent terraces
OT: Orchard terraces
AF: Agro-forestry
CT: Convertible terraces
DD: Diversion ditches
N : Normal - no need for conservation work
S : Special considerations
- : Such use not recommendable

is to use agricultural crops or pasture as a transitional means of utilizing the land until forest plantations are fully established. This has been practised mostly in national forest lands or forest reserves. The other is to bring trees and shrubs etc. into the crop or animal production systems to the benefit of both crop production and resource protection.

The latter is a permanent land use system which benefits small farmers in many ways. Woody perennials can provide fuels and fences for farmers, feed to animals, mulching and organic matter to the soils, and shade and windbreaks to crops etc. The system includes a wide range of activities with an almost endless combination of trees, agricultural crops and fodders.

Worldwide, there are many examples: Poro **(Erythrina poeppigiana)** has been grown extensively in coffee plantations in Costa Rica for shade, soil enrichment, live mulching and live fences, etc; **Albizzia spp.** have been used in tea plantations in many Asian countries. In Indonesia, Leucaena **(Leucaena leucocephala)** has been planted as contour hedges on hillsides for erosion control, soil improvement and green mulch. It is estimated that some 20 000 ha of hill land have been converted to these systems (Benge 1980). In West Africa and Rwanda many farmers use trees, fruit trees, bushes and grasses planted with agricultural crops on their farms. Many coconut plantations in the Caribbean are partly planted with bananas or used as pastures. Some small farms in Jamaica even plant coconuts, banana and citrus together.

There are tremendous possibilities and opportunities for developing new agro-forestry systems to be used by small hill farmers. However, for any system introduced to the farmers, careful consideration should be given to its effect on production as well as on erosion control. Merely introducing a row of trees to be mixed with the crops may not do any good to the farmers or the land.

With soil conservation in mind, the effectiveness of a system in controlling erosion needs to be closely examined. Wiersum (1984) recently studied surface erosion under various tropical agro-forestry systems and concluded that the key to controlling erosion in agro-forestry did not lie in the presence of the trees themselves, but rather in good management practices, either of trees or by the application of additional soil conservation technology. In fact, some agro-forestry systems are being supported by simple conservation structures. Figure 18 shows that in both the Indonesian and Rwandan systems mentioned above hillside ditches have been used to lead away excess runoff.

4.5.3 Pasture Management on Slopes

Small farmers in many developing countries use land mostly for the cultivation of food or cash crops. A small portion of a farm may be left fallow when continuous cropping is found to be unprofitable or unrewarding. Cows, draught animals, or small ruminants are usually grazed on the fallow lands. These animals are also turned loose for grazing on roadsides, stream banks or any other public land.

In fact, after land capability classification, much of the land in a farm may be classified as only suitable for pasture, i.e. gentle to steep slopes with shallow soils, stony, wet or with flooding hazards (see Appendix 3). These lands are normally unsuitable for cultivation.

In some countries such as India and Nepal there are community pastures which have been developed on steep slopes, and are under proper management.

In establishing or rejuvenating pastures on slopes or rolling land, simple conservation structures are sometimes needed to prevent erosion from heavy rains and runoff during the early stages when the land is not well covered. Hillside ditches, and contour dykes can be used. Their intervals, however, may be double or triple those used for agricultural crops. Diversion ditches are sometimes used to intercept large quantities of runoff coming from the slopes above. Mulching should be practised for reseeding pasture on steep slopes to prevent the seeds from being washed away and for a better and even germination.

Suitable fertilization or liming is often needed to help pasture establishment. Rock phosphates, if available, cost much less (Kwaengsopha 1985) and are as effective as superphosphates on soils with a low pH and high percentage Al saturation commonly found in the tropics (Sanchez 1982).

Pastures of mixed grasses and legumes usually provide a more diversified forage for various animals and maintain better fertility because of nitrogen fixation by the legumes.

Once the pasture is established, controls on grazing are necessary. The following practices are suggested which are not only good for the pastures and animals, but also for erosion control:

- Zero Grazing: In wet seasons and especially on steep slopes, zero grazing should be practised. Zero grazing is the practice of cutting the grass and stall-feeding the animals. This represents an extra labour input but children are able to do the work and the land and the animals eventually benefit.

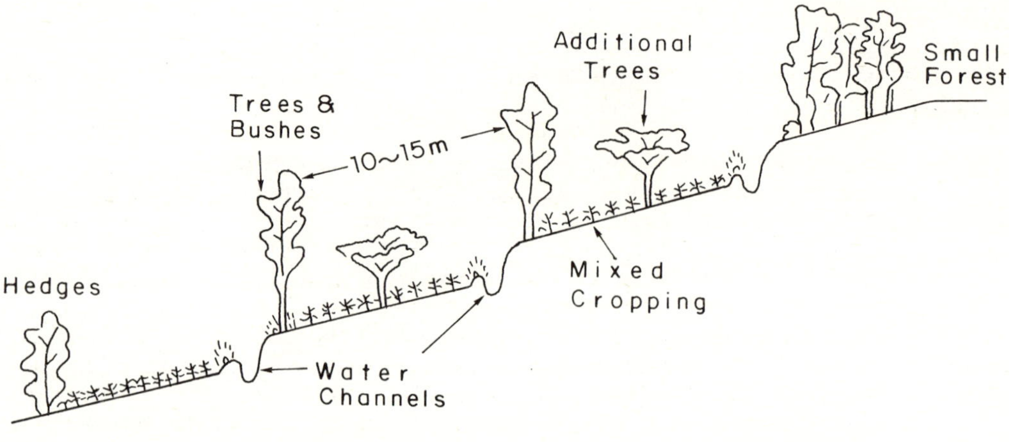

'Eco-farming' in Rwanda (after Behmel and Neumann 1981)

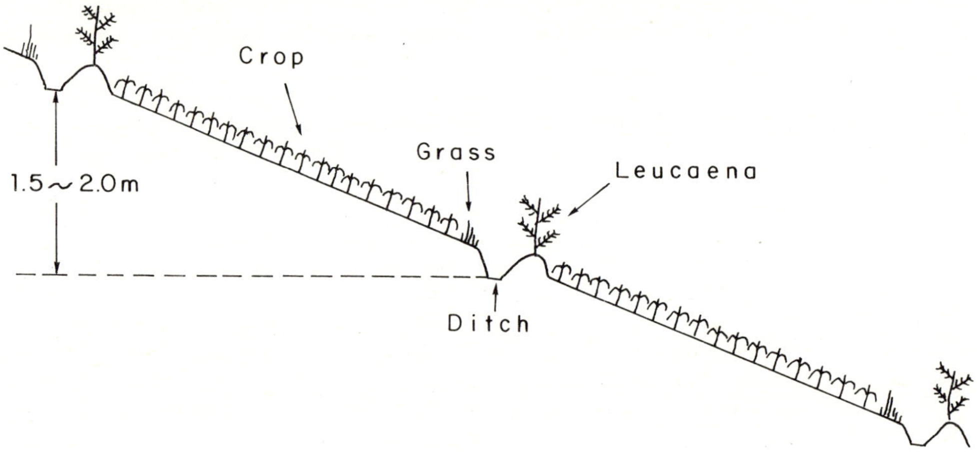

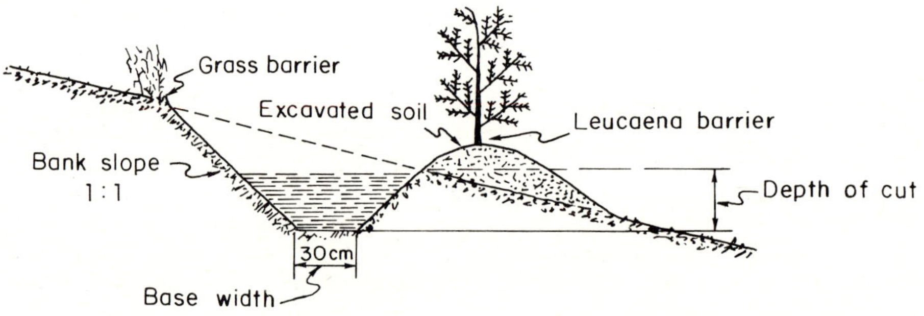

Agro-forestry system in Indonesia (after Benge 1980)

Fig. 18 Diagrams of agro-forestry systems

- Controlled Grazing: This means controlling the number of animals on a unit area. The carrying capacity of the land should not be exceeded. Supposing one hectare of grass can only support one cow, do not graze two or three unless other sources of feed are available. For community pastures certain public controls are needed.

- Rotational Grazing Is a system of pasture utilization consisting of a short period of heavy or normal stocking followed by periods of rest for plant recovery during the same season. The proper use factor of forage is about 0.65 or 0.7. When this limit is reached, the grazing should be stopped. On large community pastures, simple fences (e.g. using bamboo) can be installed to keep animals away from grazed areas. Fertilizers are needed for quick recovery. On small farms, this practice can be carried out during the zero grazing period, i.e. to control cutting so that heavily utilized areas will have a chance to recover.

5. FOLLOW-UP

5.1 MAINTENANCE

Proper maintenance is extremely important to conservation work for the future; without follow-up maintenance, soil conservation work will have no lasting benefits.

Although maintenance is essential, it is usually neglected. Small farmers who complete the land treatments and have obtained whatever financial assistance or subsidies are available tend to look for new schemes and subsidies and neglect the old work unless the maintenance is also paid for by the government. At the same time, governments are also more interested in new projects and are seldom willing to spend funds to maintain old ones. Experience shows that even field officers under pressure to accomplish annual targets tend to work in new areas and neglect checking the old ones; consequently, many soil conservation projects fail, not because of poor implementation but because of a lack of maintenance.

5.1.1 Ensuring Proper Maintenance

Drawing from experience in many developing countries, there are many ways to ensure proper maintenance of the conservation work. A brief description is given below for reference:

i. Treat the lands which the farmers can maintain

Many farmers, attracted by government incentives, and not made aware of future maintenance needs, may treat an area larger than they can maintain. Government staff should look into the farmers' capability and labour supply conditions and draw up a conservation farm plan with a practical time table for progressive treatment (see 3.5.3) and proper maintenance.

ii. Spread incentives or subsidies over several years to cover maintenance work

Governments usually expect farmers to do maintenance work themselves after incentives or subsidies have been given for the initial period of establishment. This may not be fruitful. Maintenance should be considered as an integral part of the conservation work and needs to be included in the overall incentive system. In reforestation programmes, for instance, weeding is necessary for plantation establishment after seeding or planting. Waterway and terrace maintenance are also necessary, e.g. weeding, shaping and minor repairs (see 5.1.2). Sometimes the same total of subsidies can be given to a farmer spread over 3 to 4 years (e.g. 65, 15, 10 and 10 percent respectively) to ensure good maintenance instead of in one lump sum in the first year. Any food-assisted programme can be applied in the same fashion. After several years of stabilization, many conservation structures do not need much maintenance. Also, once farmers become accustomed to maintenance, they do the work automatically.

iii. Keep treated land in use

Experience shows that if the treated land is not in use, the conservation work will not be properly maintained, regardless of the type of conservation work practised. After land treatment, government officers should assist the farmers to obtain whatever help, financial or material, they need to plant in time and put the land into productive use. When the treated land is planted to crops, the farmer will carry out daily crop management. He or she is then likely to maintain the conservation works as well, and repairs of any small breaks or other minor works will be carried out in time.

iv. Install conservation work properly from the beginning and follow up with inspections

Maintenance work can be reduced if soil conservation works, especially structures, are properly installed, including correct grade, shape, compaction, etc. Maintenance inspection is needed every year, especially for the first three years during the rainy seasons and after harvesting the crop.

v. Involve farmers with the installation of conservation works by using their own labour

It was found in many instances that if farmers were involved in the initial establishment work, they would be more responsible and interested in its maintenance.

vi. Integrate soil conservation work with other development projects

Projects such as crop expansion, irrigation, agro-industry and resettlement, etc. usually have funds for land use and cropping, and could very possibly contribute to conservation maintenance.

vii. Mobilize communities for maintaining public conservation works

Check dams, streambank protection measures, roadside conservation work, and similar structures which belong to the public or community should be maintained by people living nearby. Government needs to mobilize and organize them properly to carry out routine checks and proper maintenance, ideally with some small incentives.

5.1.2 Structure Maintenance

Newly-built bench terraces and waterways may need 20 to 30 mandays a year per hectare for maintenance. Simple conservation structures such as hillside ditches will need much less. After 3 years of stabilization, all structures generally need little maintenance, although proper shaping is always needed after cropping.

For all types of terraces, the maintenance work should include the following:

i. For benches (flat parts)

- keep the toe drains (see Figure 6) open and maintain proper gradients;
- allow all runoff to concentrate in the toe drain and ensure safe disposal to the waterway. Continuous mounds, beds or ridges on benches should always have openings to allow runoff to pass through;
- keep grass and weeds from invading the benches;
- maintain proper reverse slopes on the benches or basins and reshape them immediately after the crops have been harvested. Ploughing on terraces should be carried out carefully so that the toe drains and reverse slopes are maintained.

ii. For terrace risers

- keep grass growing on the risers and root out weeds and vines. Grass should be cut short and should not endanger the main crops;
- any small break or landslip from the riser should be repaired immediately;
- keep animals away to prevent them trampling the risers or eating the grass.

iii. For outlets (between terracs and waterways)

- check the outlets and see whether they are adequately protected by grass;
- any silt accumulated in the outlets should be cleared out immediately.

Waterways and gully structures need to be properly maintained, especially during the first rainy season or two. The needed maintenance usually includes the following:

- for grass waterways, the grass should be kept dense, short and as uniform as possible. Any tall plants and brush which weaken the grass should be removed;
- for ballasted waterways, stones should be properly fastened or keyed into the ground;
- ensure that the flow is passing over the gully structures and not going around or underneath them;
- new structures should be checked at least twice a year, once before and once after the rainy season. Any minor breaks should be repaired without delay;

5.1.3 Maintenance of Non-structural Measures

This includes the maintenance of various agronomic conservation measures mentioned previously to maintain soil structure and productivity. Generally speaking, most of this kind of maintenance can be incorporated into proper crop care and management. This does not mean, however, that non-structural measures can be spared from proper maintenance. To maintain these measures, particular attention should be given to the following:

- contour furrows or beds should be maintained with correct gradients and cross-sections throughout the cropping season so that no runoff will spill over the beds down slope;
- cover crops should be tended so that they do not hamper the growth of the main crops;
- if using mulching material, care should be taken with the replenishment or maintenance of a sufficient quantity or thickness on the ground, in addition, extra efforts must be taken with insect or pest control;
- when practising agro-forestry and using vegetative barriers, the trees or shrubs should be properly pruned to avoid over-shading and competition for moisture and nutrients with the main crops;
- multiple cropping usually needs a greater labour input per unit area in crop care, maintenance and management;
- to maintain soil productivity and structure, practices such as proper rotations, composting, green manuring, and fertilizing are still needed. Occasional ploughing or tillage of treated land is still required to create a physical soil condition, favourable for optimum crop growth.

5.2 ASSISTANCE IN CROPPING, LOANS AND MARKETING

5.2.1 Cropping

Soil conservation programmes which do not take into consideration suitable cropping will not be welcomed by small farmers. The usual question asked by small farmers after the land has been treated is: What kind of new crops, new varieties or better crops will the government help them to grow? The same type of question will be asked when a farmer is about to improve his or her pasture e.g. queries on better forage crops and animals.

It is apparent that the farmers' main objectives in participating in soil conservation programmes are more inclined towards increased production and income and less towards conservation **per se**. Experience also shows that those farmers who accept soil conservation are usually the same ones who were eager to accept a package deal of modern farming practices. Therefore to attempt to implement conservation for conservation's sake will lead nowhere and win no support from the farmers.

Assistance in cropping can be carried out either under the government's normal extension activities or under a specially-designed project. During the farm planning stage, the cropping plan for each piece of land should be discussed with the extension officer together with the soil conservation officer, and a final decision made by the farmer. The source of planting material, animals, fertilizers, chemicals, etc. should be identified, and the farmer informed. Once the farmer's land use plan is completed its conservation treatment and proper cropping should follow immediately with complete technical instructions from the extension officers. Any unnecessary waiting could mean the loss of a whole crop season. Sometimes, a soil conservation project fails, not because the work was poorly carried out, but because the cropping and production was bad. Farmers can soon lose interest.

5.2.2 Assistance with Loans

Many governments may provide loans or credit especially for soil conservation projects. Such loans may or may not include funds for crop production. Whatever the loans are for, assisting the farmer to get the needed loans on time is of vital importance.

Extension officers or soil conservation officers should be familiar with all aspects of the loans: their interest rates, grace period, limits, and payment schedules, etc. Since getting a loan will normally require certain procedures, applications with the necessary documents should be prepared well ahead of time. Unless this is done, the season for planting or other work will have passed and the farmer will blame the officer who has been trying to help.

Financial institutions that are responsible for such loans should coordinate closely with field officers of the extension and soil conservation services. If necessary, seminars for field officers and farm leaders should be held. The institutions should keep procedures flexible and be willing to solve problems from the field quickly. An effective loan supervision mechanism should be established in the field to handle daily problems.

On the other hand, farmers should be well informed of the objectives and obligations of the loans. Loans should not be diverted for other purposes and should be paid back promptly. Any defaulters may jeopardize future opportunities to obtain loans of any kind.

5.2.3 Marketing

Marketing is one of the most crucial activities in the rural areas. In many developing countries, agricultural marketing is imperfect; information is scarce; prices are unstable; and the advice given to farmers on what crops can be profitably grown is poor. Nevertheless, farmers will continue to ask what kind of crops will fetch good prices in the market.

As mentioned earlier, increase of production and especially of income may be the main motivation for farmers to participate in soil conservation programme; therefore, there is almost no option but to assist the farmers in this respect. It is futile to help farmers to produce more and better crops when there is nowhere to sell their produce, let alone profitably.

The following marketing suggestions are based on experience from developing countries, bearing in mind there are no universal answers to this complex subject.

- for small farmers, advice should be given on growing traditional crops for home consumption and for safe marketing. In the latter case, the advice should be concentrated on early producing varieties and higher quality to fetch higher prices in the existing market;

- for advanced and mid-level farmers, commercial and export crops could be introduced. The extension service should liaise with the agricultural marketing institutions or agro-industries to assure purchasing prices;

- regularly, reliable marketing information should be collected and conveyed to all the farmers in the project area so that farmers can make their own decisions on what they should grow for the coming season within the framework of the farm plan;

- in some cases, farmers can be organized to transport their own cash crops to the nearby cities for better marketing. A hill tribe in Northern Thailand was encouraged by a UN project to buy their own pick-up trucks to transport vegetables to a city. Without such an active marketing strategy, the hill tribe would not have been interested in becoming involved in better land use and settled farming;

- to install improved marketing facilities in an area, such as a grading station, or cold storage and market buildings, is also appropriate within the concept of an integrated soil conservation/watershed development project.

5.3 RECORDING AND EVALUATION

5.3.1 Keeping Conservation Farming Records

Basic records should be kept for each farm where conservation farming is practised. Some data can be entered into the farm plan (see Appendix 2) such as progress of land treatments, follow-up, cost sharing, etc. The farm plan is a record between the farmer and the government and also a blueprint to show how the farm is to be protected, improved and developed. The extension officer or the soil conservation officer in the area should keep such records with the farmer.

Some additional data may need to be collected for conservation farming records which may include cropping, major activities and returns, etc. Although it sounds complicated, the form can be of simple design to include only essential items. Appendix 5 shows an example used in Northern Thailand. Such types of records, if not practical for every farmer to keep, should at least be kept by some representative farms for further analysis. Members of youth clubs, farmer's organizations, or extension officers should assist farmers to keep such records continuously for future evaluations.

5.3.2 Acquiring Erosion and Sediment Data

Whether or not the soil conservation agency has its own experiement or hydro-meteorological network, erosion and sediment data need to be collected for further analysis. These data are very important when off-site benefits are to be evaluated. Without them the overall impact of a soil conservation project will not be known.

The usual sources for acquiring such data include the following:

- from stream gauging stations and climatic stations of the same watershed;
- from reservoir profile surveys downstream;
- from specially established soil loss and runoff plots in the area;
- from simple erosion observations and surveys in the field;
- from research results of agricultural experimental stations.

All the data should be carefully collected, analysed, kept and compared over time to get meaningful results.

5.3.3 Evaluation

Soil conservation projects are relatively difficult to evaluate because of their long-term benefits, external influences, wide distribution of benefits, etc. Nevertheless, evaluation is necessary to show the results of an investment whether the investment source is national or international. In a world where monetary resources are limited and development projects are many, governments and international financing agencies are reluctant to support continuously any project that has no serious evaluation or demonstrable long-term benefits.

The soil conservation service with the assistance of interested institutions and universities should organize an independent evaluation body to undertake periodic evaluation work. In addition, a data base and monitoring unit should be established under the responsible agency for data collection and monitoring.

In addition to farming records and erosion data mentioned in 5.3.2, specially designed surveys are needed for periodic evaluation purposes. Such periodic surveys may include the following:

- surveys of land use changes and soil conservation progress in an area or watershed using remote sensing techniques every 5 to 10 years as required;

- surveys on socio-economic conditions in the project area to compare with the baseline survey;

- surveys after major hydrologic events such as floods and long droughts and damage estimates for comparison purposes;

- other single purpose and brief surveys such as on farm income, land productivity, conservation attitude, etc. when needed.

Evaluations should be made against the original goals and expected benefits. Any major discrepancies or additional achievements and benefits should be explained. The results of evaluations should be made known to the public and government so that experience can be gained and lessons can be learnt for similar projects in the future or for other countries.

6. RESEARCH

6.1 RESEARCH NEEDS AND CHALLENGES

Agricultural research in the developing countries has been traditionally production and crop oriented. Only over the last decade have some research efforts been partly turned to environmental protection and resource conservation. However, even today, when one visits national and international agricultural research institutions, it is found that their research on soil and water conservation is still limited.

On the other hand, soil erosion and land degradation problems in developing countries are becoming more serious as population pressure increases and more marginal land is brought into food production. There are numerous problems in conserving soil and water resources in these countries, especially on steep slopes populated with subsistence farmers. Small farms as a whole have been neglected for too long, yet their erosion and runoff factors are amongst the highest in many countries.

More research is needed for developing technically sound and socially acceptable farming systems for small farmers to adopt. Whether it is hillside farming, agroforestry or any other land use systems, erosion is always a threat. The major difficulties of incorporating soil conservation into these systems lie with (i) small farmers' reluctance to accept conservation practices which may take up part of his land, and (ii) the need for simple and inexpensive, but effective, conservation measures.

6.2 RESEARCH ENVIRONMENT AND APPROACHES

In most developing countries there is a critical shortage of qualified staff, funds and equipment. With limited manpower and resources, research can only be conducted in a selective manner and with a joint effort. Some of the approaches which can be used are to:

- organize a research advisory committee among government agencies, universities and research institutes, etc. to include scientists of essential disciplines for assisting in the planning, designing and evaluation of results;

- promote research which does not need expensive instruments and sophisticated technology and so can be initiated without delay. For instance, surveys of sheet erosion with spikes and gully development with levels, identification of natural plants for conservation use, and study of farm planning techniques, are some of the basic studies which can be initiated with little difficulty;

- concentrate on problem-oriented research and on the investigations that do not need a long time to conclude;

- collect relevant research results and technical information from countries with a similar environment for reference or for local adaptation trials;

- liaise with international research institutions in the region to obtain technical assistance or to conduct joint experiments.

6.3 RESEARCH PRIORITIES AND SUBJECTS

Priorities in soil and water conservation research depend on the country's condition, resources and needs. Generally speaking, research can be divided into three categories i.e. basic, applied (technical), and managerial. Although they are all needed in a country, priorities still have to be set.

Some research subjects which need urgent attention in the field of conserving soil resources in the humid tropics are:

i. Basic research

More for international research institutions than national ones.

- Determine tolerable soil losses in the humid tropics.

- Develop simple and suitable models and equations for practical use in predicting soil loss and runoff.

- Establish the relationship between soil loss and soil productivity.

- Study erosion and runoff rates under different climates, soils, slope and land use.

ii. Applied research

Mostly for national research institutions.

- Seek low cost and effective soil conservation measures or a combination of measures for small farmers.

- Study the effects of cover crops, mulching and minimum tillage on soil properties and erosion.

- Integrate sound soil conservation practices into existing farming systems.

- Develop agro-forestry systems (including agro-pastoral) which are practical in use and effective in erosion control.

- Devise cost effective run-off disposal systems for steep slopes.

- Develop water harvesting techniques for small hill farms.

iii. Managerial research

More for national research institutions than international.

- Develop effective methods of farm planning for small farms including planning in groups or in sub-watersheds.

- Study types and need for incentives for soil conservation work and their socio-economic impact.

- Devise effective extension techniques for ensuring farmers' participation in conservation programmes.

- Study various strategies for the proper maintenance of conservation work.

The above lists could be greatly extended. The ideas selected are intended only to stimulate further thought. As, indeed, is the whole of this publication.

REFERENCES

Abujamin S. Crop residue mulch for conserving soil in uplands of Indonesia. In: Soil
1985 Erosion and Conservation, SCSA, Ankeny, Iowa.

Arnold J.E.M. Economic constraints and incentives in agro-forestry. In: Social,
1984 Economic, and Institutional Aspects of Agro-forestry. UN University, Tokyo.

Ashfaq H. The role of grasses and legumes in soil conservation practices in Northern
1979 Thailand. In: Soil Conservation Handbook, UNDP/FAO Mae Sa project, Chiang Mai, Thailand.

Behmel F. and Neumann I. Project agro-pastoral cooperation techniques, Allemmade,
1981 Nyabisindu, Rwanda. Paper presented to the workshop on Agro-forestry in the African Humid Tropics. April 1981, Ibadan, Nigeria.

Bene J.G. **et al.** Trees, food and people. International Development Research Center,
1977 Ottawa.

Benge M.D. Agro-forestation systems using **Leucaena** as erosion control barriers. USAID,
1980 Washington DC.

Bishop J.P. The dynamics of shifting cultivation, rural poor, cattle complex on marginal
1982 lands in the humid tropics. In: Social, Economic, and Institutional Aspects of Agro-forestry. UN University, Toyko.

Bochet J. Management of upland watersheds: participation of the mountain communities.
1983 Conservation Guide 8, FAO, Rome. 216 p.

Champion H.G. Report on soil erosion in the Mahogany Catchment and its control. In: The
1966 Yallahs Water Supply Scheme Feasibility Study. Min. of Overseas Develop., UK.

Chiang Y.C. Test and investigation of water storage capacity of bench terraces. J.
1965 Agri. & Fores. Vol. xiv. Taipei, Taiwan.

Council of Agriculture and MARDB. Soil conservation engineering handbook. Taipei,
1984 Taiwan.

de Graaff J. Watershed development activities proposed for Kingston Watersheds: Analysis
1981 of cost and benefits. UNDP/FAO JAM/78/006 Project Working Paper 19. Kingston, Jamaica.

FAO. Legislation principles of soil conservation. Soils Bulletin 15. Rome. 73 p.
1971

FAO. FAO-Unesco Soil Map of the World. Vol. I. Legend. Unesco, Paris.
1974

FAO. Soil conservation for developing countries. Soils Bulletin 30. Rome. 104 p.
1976a

FAO. A framework for land evaluation. Soils Bulletin 32. Rome. 79 p.
1976b

FAO. Guidelines for watershed management. Conservation Guide 1. Rome. 306 p.
1977a

FAO. Soil conservation and management in developing countries. Soils Bulletin 33. Rome.
1977b 216 p.

FAO. Forestry for local community development. Forestry Paper 7. Rome. 116 p.
1978

FAO. 1979 Watershed development, with special reference to soil and water conservation. Soils Bulletin 44. Rome. 266 p.

FAO. 1981 Agriculture: Toward 2000. Rome.

FAO. 1982 World soil charter. Rome.

FAO. 1983 Guidelines: land evaluation for rainfed agriculture. Soils Bulletin 52. Rome. 249 p.

FAO. 1984a Land, food and people. Economic and Social Development Series No. 30. Rome. 96 p. + 4 maps.

FAO. 1984b Tillage systems for soil and water conservation. Soils Bulletin 54. Rome. 296 p.

FAO. 1984c Land evaluation for forestry. Forestry Paper 48. Rome. 132 p.

FAO. 1985 Guidelines: land evaluation for irrigated agriculture. Soils Bulletin 55. Rome. 243 p.

FAO. 1988 FAO watershed management field manual - slope treatment measures and practices. Conservation Guide 13/3. Rome.

FAO/UNEP. 1983 Guidelines for the control of soil degradation. Rome. 43 p.

Greenland D.J. and Lal R. 1977 Soil erosion in the humid tropics: The need for action and the need for research. In: Soil Conservation and Management in the Humid Tropics. John Wiley & Sons, Chichester.

Gumbs F.A. **et al.** 1985 Soil erosion studies in the northern mountain range, Trinidad, under different crop and soil management. In: Soil Erosion and Conservation. SCSA, Ankeny, Iowa.

Harwood R.R. 1979 Small farm development: Understanding and improving farming systems in the humid tropics. Westview Press, Boulder, Colorado.

Hsu S.C. **et al.** 1977 Experiment on soil conservation practices for upland crops on slope-land. J. Chinese Soil and Water Conservation 8(2). Taipei, Taiwan.

Huang C.T. 1981 The rainfall erosivity index in the Southwest Area of Taiwan. J. Chinese Soil & Water Conservation 7(2). Taipei, Taiwan.

Hudson N. 1981 Soil Conservation. New edition. Batsford Academic & Educational Ltd. London.

Hudson N. 1983 Soil conservation strategies in the Third World. J. Soil and Water Conservation 38(6). SCSA, Ankeny, Iowa.

Jiang D. **et al.** 1980 Soil erosion and conservation in the Wuding River Valley. Beijing, China.

Joint Commission on Rural Reconstruction. 1977 Abstracts on soil conservation research in Taiwan. Taipei, Taiwan.

King K.F.S. 1980 Multiple-use research. Keynote address, IUFRO/MAB Conference. May 1980. Flagstaff, Arizona. USDA-FA General Technical Report WO-25. Washington DC.

Klingebiel A.A. and Montgomery P.H. 1961 Land capability classification. Agric. Handbook 210, US Department of Agriculture, Washington DC.

Kwaengsopha S. The role of improved pasture in integrated watershed management in the 1985 highlands of Northern Thailand. Paper presented to the ASEAN-US Watershed Project, Kuala Lumpur, Malaysia.

Lal R. Soil erosion and shifting agriculture. In: Shifting Cultivation and Soil 1974 Conservation in Africa. Soils Bulletin 24. FAO, Rome. 254 p.

Lal R. Role of mulching techniques in tropical soil and water management. Technical 1978 Bulletin No. 1. IITA, Ibadan, Nigeria.

Lal R. Soil erosion and its relation to productivity in tropical soils. In: Soil Erosion 1985 and Conservation. SCSA, Ankeny, Iowa.

Liao M.C. Effects of bench terraces and improved hillside ditches. J. Chinese Soil and 1976 Water Conservation 7(2) Taipei, Taiwan.

Lowdermilk W.C. Conquest of the land through 7,000 years. USDA Agric. Information Bull. 1953 99. Washington DC.

Mannering J. V. and Fenster C.R. What is conservation tillage? J. Soil and Water 1983 Conservation 38(3). SCSA, Ankeny, Iowa.

Manshard W. Tropical Agriculture. Longman, London. 1974

Marston D. **et al.** Soil conservation and land development in Northern Thailand. In: Soil 1985 Erosion and Conservation. SCSA, Ankeny, Iowa.

Moormann F.R. and Van Wambeke A. The soils of lowland rainy tropical climates: Their 1978 inherent limitations for food production and related climate restraints. In: Plenary Papers. Vol. 2. 11th Inter. Congr. of Soil Science. Uni. of Alberta, Edmonton, Canada.

Morgan R.P.C. Priorities for technical research in soil conservation. Paper presented 1984 to the International Workshop on Soil Erosion and its Countermeasures. Nov. 1984. Chiang Mai, Thailand.

Nair P.K.R. The re-marriage of crops and trees. International Agric. Development. July/ 1983 August 1983.

Ofori C.S. Shifting cultivation: Reasons underlying its practice. In: Shifting culti- 1974 vation and soil conservation in Africa. Soils Bulletin 24. FAO, Rome. 254 p.

Ogborn J. No-till cropping systems for smallholder farmers. Appropriate Technology 9(2). 1982

Ong S. E. Small farm development in Asia: An overview of its problems and perspectives. 1982 East-West Center, Honolulu, Hawaii.

Osuji G.E. and Babalola O. Tillage practices on a tropical soil. J. Environment Manage- 1982 ment 14. London.

Posner J.L. Cropping systems and soil conservation in tropical America. Rockefeller 1981 Foundation. New York.

Posner J.L. and McPherson M.F. The steep sloped areas of tropical America: Current 1981 situation and prospects for the year 2000. Rockefeller Foundation. New York.

Sanchez P.A. Properties and management of soils in the tropics. John Wiley & Sons, 1976 Chichester.

Sanchez P.A. A legume-based, pasture production strategy for acid infertile soils of 1982 tropical America. In: Soil Erosion and Conservation in the Tropics. ASA special public. No. 43. Madison, Wisconsin.

Sfeir-Younis A. Economic aspects of soil conservation programs in less-developed
1983 countries (LDC's). Water International. Vol. 8. Elsevier Sequoia, Netherlands.

Shaner W.W. **et al.** Farming systems research and development: Guidelines for developing
1982 countries. Westview Press, Boulder, Colorado.

Sheng T.C. A treatment-oriented land capability classification scheme, for hilly
1972 marginal lands in the humid tropics. Paper presented to the Latin American Watershed Management Seminar. Nov.-Dec. 1971. La Plata, Argentina. In: Report No. TA 3112. FAO, Rome.

Sheng T.C. Physical planning for watersheds. UNDP/FAO JAM/78/006 Project Working Paper
1981a 17. FAO, Rome.

Sheng T.C. The need for soil conservation structures for steep cultivated slopes in the
1981b humid tropics. In: Tropical Agricultural Hydrology. John Wiley & Sons, Chichester.

Sheng T.C. Erosion problems associated with cultivation in humid tropical hilly regions.
1982 In: Soil Erosion and Conservation in the Tropics. ASA Special Public. No. 43. Madison, Wisconsin.

Sheng T.C. and Michaelsen T. Runoff and soil loss studies in yellow yams. UNDP/FAO JAM/
1973 505 Project Working Document. Kingston, Jamaica.

Sheng T.C. **et al.** The effects of different structures on erosion and runoff. Paper
1981 presented to the Southeast Asia Regional Symposium on Problems of soil erosion and sedimentation. Jan. 1981. Bangkok, Thailand.

Tautscher O. Torrent and erosion control. UNDP/FAO PHI/72/006 Project Report. Manila,
1975 The Philippines.

ter Kuile C.H.H. Research and shifting cultivation at the International Institute of
1984 Tropical Agriculture. In: The Future of Shifting Cultivation in Africa and the Task of Universities. FAO, Rome. 198 p.

UN. United Nations Conference on Desertification: Round-up Plan of Action and Resolu-
1878 tions. New York.

Vine H. Experiments on the maintenance of soil fertility at Ibadan, Nigeria, 1922-1951.
1953 J. Exper. Agric. 21.

von Uexhull H.R. Managing Acrisols in the humid tropics. Exten. Bulletin No. 22. Food
1984 and Fertilizer Technology Center, Taipei, Taiwan.

Wichaidit W. **et al.** Forests, forest development, shifting cultivation and erosion in
1977 Northern Thailand. NADC, Chiang Mai, Thailand.

Wiersum K.F. Surface erosion under various tropical agro-forestry systems. Paper
1984 presented to the IUFRO/EWC Symposium on the Effects of Forest Land Use on Erosion and Slope Stability. May 1984. Honolulu, Hawaii.

World Bank. World Development Report. Washington DC.
1983

Wu H.L. A review of soil conservation measures on slopeland watershed in Taiwan. MS
1986 Thesis, Utah State University, Logan, Utah.

APPENDIX 1

EXAMPLE OF A QUESTIONNAIRE FOR CONSERVATION FARM PLANNING

Part I: Interviewing (Material to be treated as Confidential)

Village: Date of Visit:

House No: Enumerators:

1.1 Name of Farmer (Head):

1.2 Age: Education:

1.3 Sex:

1.4 Location of Farm:

1.5 Area and Ownership (By parcels): Total area:....................

Within the village		Outside the valley or swidden
Area	Ownership*	Area
1)		1)
2)		2)
3)		3)
4)		4)
5)		5)

* owned, leased or operated

1.6 Present land use:

1) Main crop..............ha
2) Vegetables.............ha
3) Fruit trees............ha
4) Pastures/Fallow..........ha
5) Fallow...................ha
6) Wood-lots................ha
7) Others...................ha

Explain the reasons of Fallow and lands not being used:

1.7 Living:

1) On the Farm (   ) or not on the Farm (   )
2) Partly on the Farm (   )
3) Produces sufficient rice for home consumption (   )
4) Rice is not self-sufficient (   )

1.8 Family: Married ( ) Single ( ) Separated ( )

1) No. of children............
2) No. of Family members available for work ....

1.9 Labour Supply:

1) All from family members ( )
2) Partly depending on hired labourers ( )
3) Partly on co-operative basis ( )
4) Others (specify)
5) Do you expect a hand tractor will become necessary in the near future? Yes/No

1.10 Total annual income from Farm:

1) From farm in the village.........................
2) From land outside the village....................
3) Other sources..................................

Total:

1.11 Property:

1) House....................
2) Small stock.............
3) Cattle..................
4) Any other................

1.12 Farmer's proposed plan for the next 3 years:

Land use changes (+ or -)

1) Main crops............ha
2) Vegetables............ha
3) Fruits................ha
4) Pastures..............ha
5) Fallow................ha

Infrastructure

1) Road.....................
2) Irrigation...............
3) Buildings................
4) Domestic water supply.....
5) Others (specify)..........

Animal Husbandry

6) Small stock...........
7) Cattle.................

1.13 Farmers interest in conservation measures: Yes/No

If yes, specify the types

1)
2)
3)
4)
5)

If no, specify the reasons

1)
2)
3)
4)
5)

1.14 Existing area with conservation measures..................ha (including paddies)

Proposed area with conservation measures..................ha

1.15 Type of incentive required to undertake Conservation farming: (select 3 items)

1) Seed and plant materials
2) Fertilizers and pesticides
3) Assistance in labour
4) Technical supervision
5) Cash
6) Irrigation
7) Marketing and storage
8) Service of mechanisation
9) Others (specify)

1.16 Remarks:

1) How many of your family members work with the Project?

2) Marketing facilities

a) For agricultural commodities

b) Cooperative basis or others

c) Transportation by vehicle

Part II: Planning

2.1 Land:

Parcel	Area	Slope	Soil Depth	Land Capability	Proposed Land use	Conservation Needs
Within the village						
1)						
2)						
3)						
4)						
5)						
Outside the valley or swidden						
1)						
2)						
3)						
4)						
5)						

2.2 Water:

1) Available for Irrigation: Yes/No

2) To be developed: Yes/No

How?

2.3 Road:

1) Extra road needs: Yes/No

2) Type road needed:

2.4 Tools:

For conservation treatment:

1) Hand tools ( )

2) Animal draught tools ( )

3) Machines ( )

2.5 Labour:

1) No. of people who will perticipate in land treatment:

2) Source of labourers

2.6 Others: Attach a sketch map showing all the parcels, etc.

APPENDIX 2

EXAMPLE OF A FARM PLAN FOR CONSERVATION AND DEVELOPMENT FROM JAMAICA*

Farm Plan No. *0019* Hectares: *1.5*

Farmer's Name: *George Henry* Age: *39*

Address and District: *Windsor Forest*

Specific location of farm: *Intersection to Ramble from Windsor Forest*

I hereby agree to undertake the execution of the work specified in the fore going farm plan to the entire satisfaction of the JAM/78/006-GCP/NOR project and in the case of any disagreement with respect to the execution of the plan to abide by the decision of the Permanent Secretary in the Ministry of Agriculture.

I further agree to allow any duly appointed Officer of the project to enter my holding at any reasonable time for the purpose of assisting or inspecting the work being carried out or to be carried out.

I agree to maintain all soil conservation treatments and other improvements in good condition. It is agreed the Project shall not be held liable for any damage that may result from construction and use of works or improvement.

Sgd:	*George Henry* (Farmer)	*15 September 1981* Date
Submitted:	*W. Wellington* (Extension Officer)	*16 September 1981* Date
Recommended:	*L. Latty* (Senior Soil Co. Officer)	*16 September 1981* Date
Recommended:	*R. Roach* (Senior Extension Officer)	*22 September 1981* Date
Recommended:	*B. Cameron* (Senior Extension Officer)	*28 September 1981* Date
Approved:	*D.M. Ramsay* (Project Director)	*28 September 1981* Date

---

Parcels operated	Size	Owned	Leased	Rented	Familyland	Title Farmed: Partially	Years	Location
This parcel (1)	*1.5 ha*		*Leased*				*7 yrs*	*Windsor Forest*
No. (2)	-							
No. (3)	-							

* All measurements converted to metric units.

Major crops on other parcels: *Banana, plantain and gungo peas*

Name of Farm owner address: *Windsor Forest Dist., Pine Head P.A. St. Thomas*

Does owner approve of this farm plan? *Yes*

Name of Spouse: *Miss Gloria Clarke* Does spouse know about the plan? *Yes*

No. & age of dependents at home: *five children*

Dependents and ages living away: *2 dependents of age 20 and 24, both spouse's*

Who does work on the farm: *Mixed labourers and farmer*

Amount of labour hired (man-days): *2 men during cropping and reaping*

Livestock now kept (kind & No.): *None*

Other sources of income: *Plumbing and grocery shop (by spouse)*

Will a loan be needed? *Yes* Approximate amount: *Not specified*

Major items needing loan: *It is needed for a water tank since sometimes municipal water is not available*

ADDITIONAL FARM INFORMATION

1. Record of changes in land use:

Change	Hectares Planned	Hectares Applied
(a) Additional land put into cultivation or into permanent crops:	*0.84 (from the ruinate)*	*0.84*
(b) Cropland over 25 or FT P. or F. Cap class taken out of cultivation	*none*	*none*

2. Description and special soil erosion problems existing on the farm.

Sheet erosion. Farmer also recognized gully erosion.

3. Farmer(s) opinion as to whether farm crop yields the past three years have been decreasing ________ or increasing *X* ________( )

Explain: *It is because the farmer increases his input each year, such as the use of fertilizer, sprinkler irrigation, etc.*

4. Describe the present livestock operation: *Not applicable*

5. Describe the condition and productivity of present crops and possible diagnosis of problems: *It is quite reasonable with middle level crop management*

6. Describe the family house if requesting a new house or improvement:

It has 7 rooms completed, 3 uncompleted and it is in fair condition

NOTES ON PLANNING AND FOLLOW-UP

Narrate pertinent information that will affect the application of this plan. Also note items needing later follow-up assistance of a special nature.

DATE	NOTES
17/11/81	*Terraces lined out*
23/11/81	*Bench terraces construction begins*
1/12/81	*Waterway lined out*
3/12/81	*Bench terraces checked*

19/1/82	Banana establishment begins as shade for coffee
5/2/82	Bench terraces checked
19/2/82	Farmer wanted to plant citrus trees on the coffee area above orchard terraces. It is submitted for consideration
25/2/82	Poultry manure received
1/3/82	Coffee received. Planting to begin immediately
3/3/82	Hillside ditch checked
29/4/82	Waterway to be completed

PRESENT LAND USE

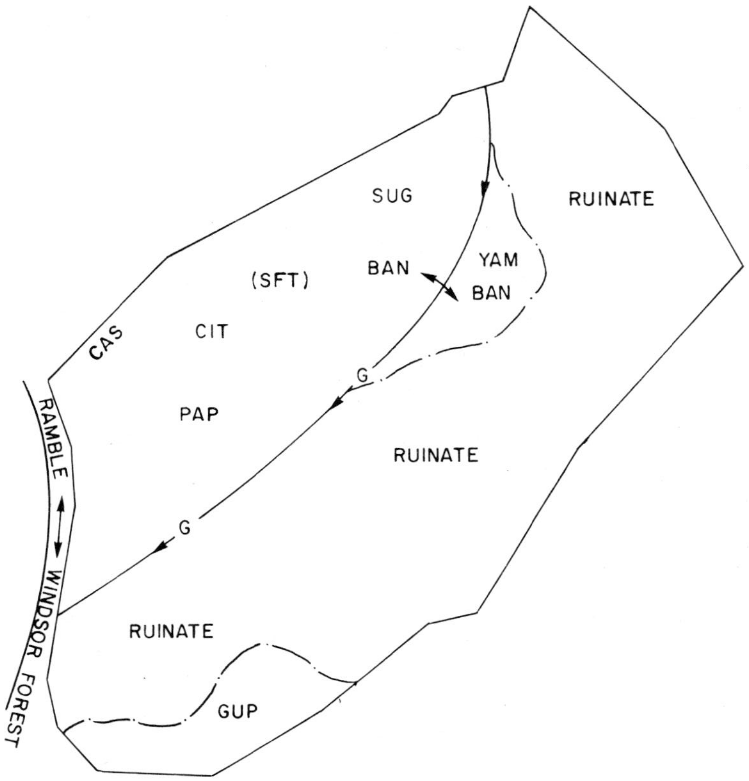

SCALE :- 1 : 1200 AREA = 1·5 ha

LAND CAPABILITY AND SOILS

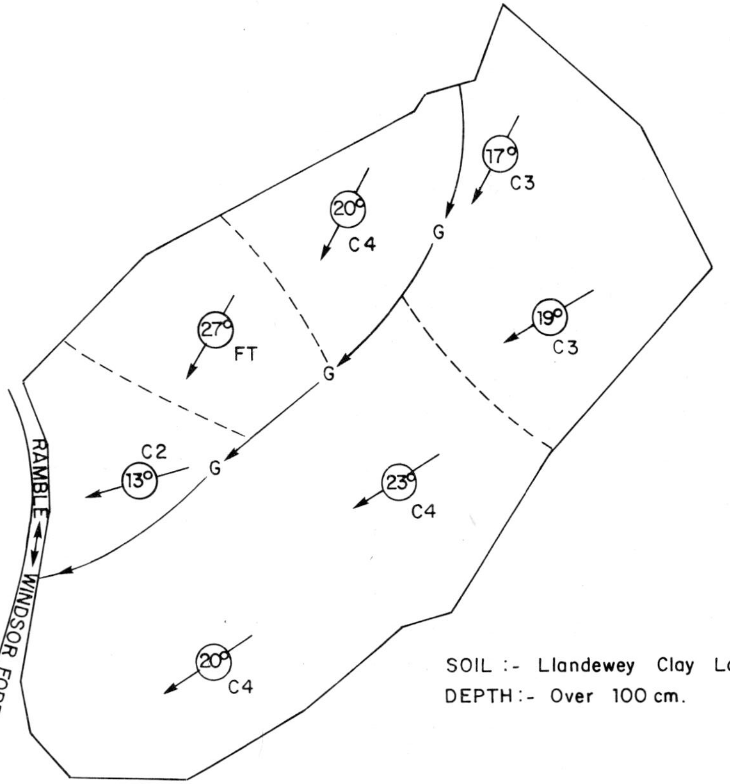

SOIL :- Llandewey Clay Loam
DEPTH :- Over 100 cm.

Farm No 04/132

Farmer's Name: George Henry

PROPOSED DEVELOPMENT

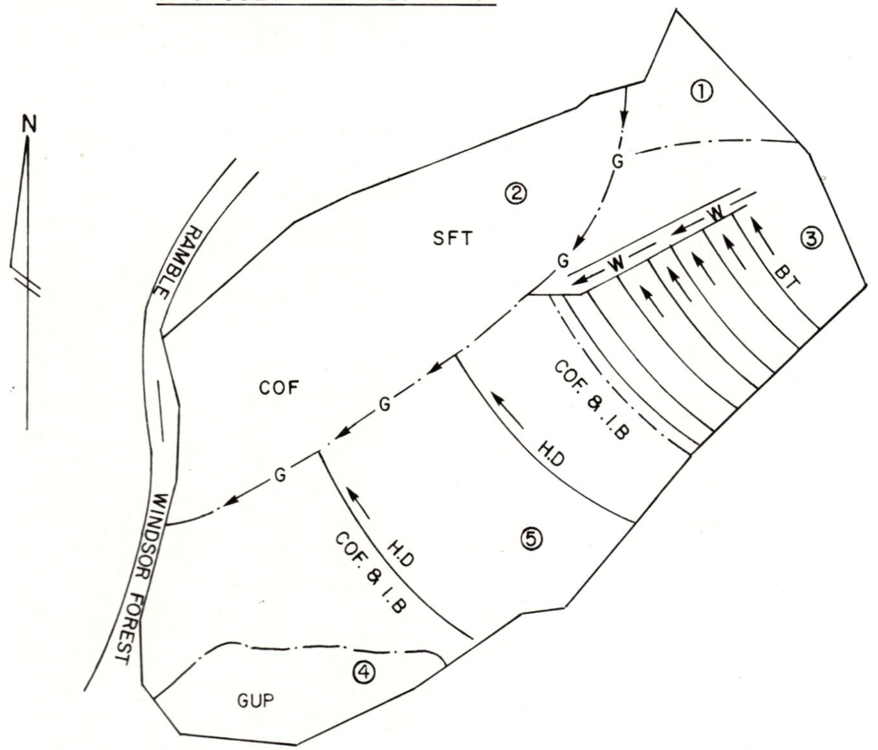

SCALE :- 1:1200

LEGEND

═══	Main Road
───	Farm Boundary
—·—·—	Parcel Boundary
①	Parcel Number
—(27°)—	Slope & General Direction
- - - - - - -	Delineation of Slope Boundary
BT	Bench Terrace
HD	Hillside Ditch
IB	Individual Basin
$C_1 - C_4$ . .	Cultivable Land
FT	Fruit or Food Trees
—W—W—	Waterway
— G —	Natural Gully (Waterway)
───	Slope of Structures
BAN	Banana
CAS	Cassava
CIT	Citrus
COF	Coffee
GUP	Gungo Peas
PAP	Papaw
SFT	Scattered Fruit Trees
SUG	Sugar Cane

CONSERVATION AND DEVELOPMENT FARM PLAN

Farm No. *04/132*

Farmer's Name *George Henry*

Farm Plan No. *0019*

Area No.	Hectares	Planned land use	Treatment and development practices and specifications planned	Amount planned	Average cost rate	Total cost $	Project cost $	Farmer cost $	Date to do and amount			Total
1	*0.07*	*Hut*										
2	*0.46*	*FT (scattered)*	*Periodic maintenance by farmer as necessary*									
3	*0.33*	*BT*	*BT 13' - 20°*	*300 m*	*$ 4.2/m*	*1260*	*945*	*315*				
		Vegetables	*Onions*									
			Poultry manure	*250 bags*	*$ 1/bag*	*250*	*187*	*63*				
			Fertilizers	*(not for first crop)*								
4	*0.10*	*GUP*	*Periodic maintenance by farmer as necessary*									
5	*0.42*	*COF and IB*	*Seedlings*	*1 110*	*20¢/plant*	*222*	*222*	*(Totally project contribution)*				
			Holes and IB		*67¢/plant*	*740*	*555*	*185*				
			Poultry manure	*220 bags*	*$ 1/bag*	*220*	*165*	*55*				
			Planting and fertilizers		*30¢/plant*	*333*		*333*	*(Totally farmers' contribution)*			
		HD	*FT to be planted*	*120 m*	*$ 2.7/m*	*324*	*240*	*84*				
Total	*1.51*					*3349*	*2314*	*1035*				

PROGRESS AND ACCOMPLISHMENT RECORD

FARM PLAN NO: ____________________ FARMER'S NAME: *GEORGE HENRY*

Date	Treatment of Practice	Voucher Number	Amount Planned Quantity	Rate $	Amount Completed Quantity	Amount	Amount Remaining	Quality Amount Remaining
30/11/81-4/12/81	BT	By contract	300 m	63/20 m	40 m	189.00	240 m	
14/12/81-22/12/81	BT		–	"	50 m	157.50	190 m	
22/12/81-20/1/82	BT		–	"	90 m	283.50	100 m	
2/2/82-15/2/82	BT		–	"	40 m	126.00	60 m	
19/2/82-11/3/82	HD		120 m	40/20 m	120 m	240.00	–	
19/2/82-11/3/82	BT		–	63/20 m	60 m	189.00	–	
3/3/82	COF		1110	50c/ basin		555.00		

APPENDIX 3

A SCHEME OF LAND CAPABILITY CLASSIFICATION FOR JAMAICA

A Treatment-oriented Scheme especially for Hilly Watersheds

Slope / Soil depth	Gently sloping < 7°	Moderately sloping 7°-15°	Strongly sloping 15°-20°	Very strongly 20°-25°	Steep 25°-30°	Very steep > 30°
Deep (D) > 36 in (90 cm)	C_1	C_2	C_3	C_4	FT	F
Moderately deep (MD) 20-36 in (50-90 cm)	C_1	C_2	C_3	C_4 / P	FT / F*	F
Shallow (S) 8-20 in (20-50 cm)	C_1	C_2 / P	C_3 / P	P	F*	F
Very shallow (VS) < 8 in (20 cm)	C_1 / P	P	P	P	F	F

* Forest land, or properly designed agro-forestry.

Notes

1. Symbols for most intensive tillage or uses:

C_1 : Cultivable land 1, up to 7° slope, requiring no, or few, intensive conservation measures, e.g. contour cultivation, strip cropping, vegetative barrier, rock barrier and in larger farms, broadbase terraces.

C_2 : Cultivable land 2, on slopes between 7° and 15°, with moderately deep soils needing more intensive conservation e.g. bench terracing, hexagon, miniconvertible terracing for the convenience of four-wheel tractor farming. The conservation treatments can be done by medium-sized machines such as Bulldozer D5 or D6.

C_3 : Cultivable land 3, 15° to 20°, needing bench terracing, hexagons and miniconvertible terracing on deep soil and hillside ditching, individual basin on less deep soil. Mechanization is limited to small tractor or walking tractor because of the steepness of the slope. Terracing can be done by a small tractor with 8 ft wide blade.

C_4 : Cultivable land 4, 20° to 25°, all the necessary treatments are likely to be done by manual labour. Cultivation is to be practised by walking tractor and hand labour.

P : Pasture, improved and managed. Where the slope is approaching 25°, and when the land is too wet, zero grazing should be practised. Rotational grazing is recommended for all kinds of slopes.

FT : Food trees or fruit trees. On slopes of 25° to 30°, orchard terracing is the main treatment supplemented with contour planting, diversion ditching and mulching. Because of steepness of the slopes, interspaces should be kept in permanent grass cover.

F : Forest land, slopes over 30°, or over 25° where the soil is too shallow for any of the above soil conservation treatments.

2. Any land which is too wet, occasionally flooded or too stony which prevents tillage and treatment should be classified as: (a) below 25° slope : pasture; (b) above 25° : forest.

3. Gully dissected lands which prevent normal tillage activities should be classified as: forest/pasture.

4. Mapping Symbols: Could be labelled as follows:

$$\frac{\text{Most intensive use}}{\text{soil - slope - depth}}$$

example: $\frac{C2}{32 - 2 - D}$

means: $\frac{\text{Cultivable Land 2}}{\text{Wirefence Clay Loam - } 7^\circ \text{ to } 15^\circ \text{ - 100 cm}}$

or, could be simply labelled as C_2.

(a) Slope classification

Slopes are divided into six categories, each having its implications for conservation treatments and the kind of tools to be used:

<7° Flat to gently sloping. Broadbase terraces or other simple conservation treatments can be used up to 7°. Full mechanization for cultivation is applicable in this category. This slope class may not be common in hilly marginal zones.

7°-15° Moderately sloping. Medium-sized machines such as a Bulldozer D5 or D6 can be employed for bench terracing. Four wheel tractor mechanization for cultivation can be applied.

15°-20° Strongly sloping. Small-sized machines such as D4 can be employed for conservation treatments. Small tractors, or walking tractors can be used for cultivation.

20°-25° Very strongly sloping. Manual labour for conservation treatments. Hand labour and walking tractor for cultivation.

25°-30° Steep. Only for permanent tree crops such as food trees, fruit trees or forest. Manual labour for treatments.

>30° Very steep. Needs forest cover.

(b) Soil depth

Soil depth is divided into four classes. Here the depth refers to the effective depth of the soil which machine or manual labour can cut for conservation treatments and which plant roots can penetrate.

< 20 cm Very shallow. Only on nearly level land can cultivation be practised.

20-50 cm Shallow. Only below 20° slopes can this be cultivated with conservation treatments.

50-90 cm — Moderately deep. A 25° slope, for instance, needs about 90 cm of soil to make narrow terraces 2.4 m wide.

> 90 cm — Deep. No further soil depth classification is needed because the riser or terrace is limited to 1.8 m height which is 90 cm cut and 90 cm fill.

(c) Other limiting factors

Land which is too wet, has poor drainage, occasionally floods or is too stony, which permanently limits the tillage or treatment, should be classified for lower or less intensive uses. On slopes under 25° such land can be used as pasture, whereas on slopes over 25° forest cover is ideal (so far as erosion control is concerned). Gully-dissected land which prevents any tillage activity should be put under permanent cover.

(d) Capability classes

Land is classified according to the range and intensity of uses which are possible without incurring land degradation. There are four major classes - cultivable land, pasture, food trees and forest [1].

Only cultivable land has four sub-classes, each having implications for needed conservation treatments and tools to be employed. Use according to or within the capability class is encouraged, whereas use beyond the capability class is discouraged.

(e) Soil conservation treatment

In addition to the most popular conservation treatments on gentle slopes (below 7°) such as broadbase terraces and strip cropping etc., six major treatments for steeper slopes are taken into account for the basis of this classification scheme. These six treatments, which have been established in the hill slopes of Taiwan as well as in the western part of Jamaica are particularly suited for the humid tropics. Bench terraces, hillside ditches and individual basins can be used to treat slopes up to 25° if the soils are deep enough. Orchard terracing can be applied on 25° to 30° slopes. Mini-convertible terracing and hexagons for full mechanization are to be employed on slopes up to 20°. All of them are mainly reverse sloped terraces of varying widths.

---

[1] The term is used in its broad sense of forest cover, including natural, man-made, economical or protective.

APPENDIX 4A

EXAMPLE OF CURRICULUM FOR EXTENSION OFFICERS TRAINING COURSE
(From Jamaica) (3 Weeks)

No.	Course Subjects	Hours: Lectures	Hours: Field and Lab Practices	Remarks
1	Soil Conservation Programmes in Jamaica	1	-	
2	Land Use and Soil Conservation	2	-	
3	Concepts of Watershed Management	2	-	
4	Forestry and Watershed Management	2	-	
5	Soil Erosion	1	1	
6	Land Capability Classification	2	2	
7	Fundamental Surveying	2	6	
8	Use and Care of Instruments	1	-	
9	Maps, Map Reading & Area Calculation	1	2	
10	Bench Terracing & Maintenance	3	5 (items 10–11 combined)	Construction practices
11	Simple Terracing Systems	2		
12	Using of Specification Tables	1	1	
13	Run-off, Waterways & Gully Erosion Control	3	5	
14	Road Erosion Control	2	-	
15	Soil and Fertility Management	2	3 (items 15–18 combined)	Field observations combined
16	Agro-forestry	2		
17	Agronomic Conservation Measures	2		
18	Grass & Legume Establishment	2		
19	Farm Planning	2	4	
20	Socio-Economic Implications of Soil Conservation Programmes	2	-	
21	Soil Conservation Extension	3	-	
22	Field Study Tours	-	12	
	Total	40	41	

Note:

(1) One day comprises 6 working hours.
(2) Full training notes are given to the participants.
(3) Films and slide shows, orientations, evaluation etc. are not included.

APPENDIX 4B

EXAMPLE OF A CURRICULUM FOR BASIC TRAINING COURSE IN WATERSHED MANAGEMENT AND SOIL CONSERVATION (FROM JAMAICA)
(4 Weeks)

No.	Course Subjects	Hours: Lecturing	Hours: Field and Lab Practices	Remarks
1	Soil Conservation Programmes	1	-	
2	Concepts of Watershed Managment	2	-	
3	Soil Conservation Philosophy	2	-	
4	Introduction to Hydrology	2	-	
5	Principles of Hydraulics	2	-	
6	Watershed Behaviour	2	-	
7	Soil Erosion	2	2	
8	Airphoto-Interpretation	2	5	
9	Physical Survey of Watersheds	2	-	
10	Social and Economic Aspects in Watershed Survey	2	-	
11	Maps, Map Reading & Area Calculations	1	1	
12	Fundamental Surveying	2	5	
13	Care of Instruments	1	-	
14	Levelling and Contouring	3	7	
15	Land Capability Classification	2	5	
16	Bench Terracing and Maintenance	3	14 (nos. 16–18 combined)	Field work combined and conducted after No. 19 staking and layout.
17	Hillside Ditching, Individual Basins and Orchard Terracing	2		
18	Mini-Convertible Terraces, Hexagons and Other Conservation Treatments	2		
19	Staking and Layout of Conservation Treatments	2	7	
20	Runoff Water	2	-	
21	Construction of Small Structures	2	7 (nos. 21–23 combined)	Field work combined.
22	Waterways	2		
23	Gully Erosion Control	2		

24	Road Erosion Control	2	5	
25	Stream Bank Erosion Control	2	-	
26	Agronomic Conservation Measures	2	-2	Combined field studies
27	Grass & Legume Establishment	2		
28	Soil Fertility Management	2	-4	Combined field work
29	Multiple Cropping	2		
30	Agro-forestry and Forestry	2	2	
31	Minor Irrigation	2	2	
32	Incentives for Small Farmers	2	-	
33	Soil Conservation Extension	2	-	
34	Conservation Farm Planning	2	5	
	Total	67	68	

Note:

(1) Ten (10) hours of surveying course (No. 11, 12 & 13) can be eliminated if participants already have sufficient training and experience elsewhere.

(2) Full lecture notes are given to the participants at the beginning of the course.

(3) One day comprises 7 working hours.

(4) Films and slide shows, discussion and review sessions, evaluations and examination etc., are not included.

(5) The following subjects were included in one of the previous courses:

1.	Government Policy on Small Farmers	1 hour
2.	Watershed Protection in Jamaica	2 hours
3.	Water Resources in Jamaica	2 hours
4.	Agricultural Planning Related to Hilly Watersheds	1 hour
5.	Geology and Landforms in Jamaica	2 hours
6.	Rainfall in Jamaica	2 hours
7.	Major Soils in the Uplands of Jamaica	2 hours
8.	Farm Management	2 hours
9.	Mechanization on Hilly Lands	2 hours
10.	Tree Crops in Watershed Management	2 hours

APPENDIX 4C

EXAMPLE OF A CURRICULUM FOR PRACTICAL TRAINING IN SOIL CONSERVATION

(2 Weeks)

No.	Course Subjects	Hours: Lectures	Hours: Field and Lab Practices	Remarks
1	Soil Conservation & Watershed Management	1	-	
2	Introduction of Land Capability Classification	2	-	
3	Introduction of Hand Level and Dumpy Level	2	6	
4	Contouring and Levelling	1	7	
5	Bench Terracing and Other Treatments	2	-	
6	Layout and Staking of Bench Terraces and Hillside Ditches, etc.	1	6	
7	Construction and Maintenance of Bench Terraces and Other Treatments	1	9	
8	Introduction of Waterways	2	-	
9	Waterway Profile Survey	-	3	
10	Waterway Construction	1	7	
11	Road Erosion Control	1	3	
12	Use of Compass	1	5	
13	Map Reading	1	2	
14	Map Making	1	2	
15	Slope Analysis	1	1	
16	Area Calculation	1	1	
	Total	19	52	

Note:

(1) Outlines of above subjects are given to the participants, supplemented by some full lecture notes.

(2) The above list does not include the courses for evaluation and orientation, etc.

(3) One day comprises 8 working hours.

APPENDIX 5

EXAMPLE OF A CONSERVATION FARMING RECORD SHEET

Farmer:

Record Keeper:

Field No:

Conservation Treatment:

Area :

Cropping System : 1. 3.

2.

Seed Sources : 1.

: 2.

: 3.

Culture method :

Crop No: Area m^2

1. Land preparation: m/d (date:)
2. Amount of seed required: kg
3. Date of planting: 1) Proposed date: ....

   2) Actual date : ....

   3) Labour : ....m/d

4. Weeding 1st: m/d(Date)

   2nd: m/d (Date)

   3rd: m/d (Date)

5. Fertilizer: Time, kind and amount: 1st m/d (Date)

   2nd m/d (Date)

   3rd m/d (Date)

6. Insect and pest control: (time, type and amount applied)

   1st: m/d (Date)

   2nd: m/d (Date)

   3rd: m/d (Date)

7. Harvest: (Including cutting, threshing, drying storage etc.)

   1) Date of harvest:
   2) Labour : m/d
   3) Yield : kg

8. Remarks: (marketing price, etc.)